Something Went Wrong

ASSISTED BY: Mieko Anders, Rachel Z. Arndt, Amy Barnes, Landon Bates, Ellie Bozmarova, Rita Bullwinkel, Jessica DeCamp, Tasha Kelter, Trang Luu, Jane Marchant, Christina Ortega, Lydia Oxenham, Leon Pan, Alexandre Pomar, Enzo Scavone, Courtney Soliday. WEB DEVELOPMENT: Brian Christian. PUBLISHING ASSOCIATE: Eric Cromie. COPY EDITOR: Daniel Levin Becker. FOUNDING EDITOR: Dave Eggers. EXECUTIVE AND EDITORIAL DIRECTOR: Kristina Kearns. ART DIRECTION: Sunra Thompson. MANAGING EDITOR: Claire Boyle.

JACKET AND INTERIOR ILLUSTRATIONS: Sunra Thompson.

SIDEBARS COMPILED BY: Landon Bates, Trang Luu, Leon Pan.

EDITORIAL ADVISORS: The Electronic Frontier Foundation, with special thanks to Bill Budington, Cindy Cohn, Andrew Crocker, Bennett Cyphers, Hugh D'Andrade, Cory Doctorow, Elliot Harmon, Aaron Jue, Jason Kelley, Dave Maass, Corynne McSherry, Soraya Okuda, Seth Schoen.

MCSWEENEY'S INAUGURAL PUBLISHING BOARD: Natasha Boas, Kyle Bruck, Carol Davis, Brian Dice (PRESIDENT), Isabel Duffy-Pinner, Dave Eggers, Caterina Fake, Hilary Kivitz, Nion McEvoy, Gina Pell, Jeremy Radcliffe, Jed Repko, Julia Slavin, Vendela Vida.

Printed in China

THE END OF TRUST

McSWEENEY'S

SAN FRANCISCO

We recently learned about a function deep in the cobwebby bowels of the iPhone. Getting there requires a compass and sensible hiking shoes. From the SETTINGS menu, we opened the PRIVACY tab, went to LOCATION SERVICES, and scrolled all the way down to SYSTEM SERVICES: there lay the entrance to the SIGNIFICANT LOCATIONS log. We entered our passcode and we were in. There it was: each trip into the office, complete with time stamps of our arrival and departure. There it was: the bodega where we stopped during the Families Belong Together march. There they were: the clandestine trips to the Crepe House, our guilty-pleasure breakfast spot, the one we only patronize alone, hunched over, hoping no one sees us there. But someone did–Apple saw us there. Apple saw us there on August 3 at 11:13 a.m. Apple saw us there on June 11 at 10:44 a.m. Apple saw us there on January 5 at 1:43 p.m.

These aren't just embarrassing secrets laid bare–these are the details of our whole life, mapped out and catalogued. The people we spend our time with, the meetings we attend, our penchant for crepes during work hours. Innocence lost, we tapped through the thorny brambles of our PRIVACY options to the beat of our quickened pulse. We realized that we'd become so used to breezing past user agreements that we'd forgotten privacy was something we should still expect, or rather demand. If we're being honest, we'd hand over all of our Facebook contacts just to watch the final episode of *MasterChef*

Junior for free. But every day that we ignore the consequences of these little trade-offs, we're tacitly consenting.

Through this collection, our first-ever entirely nonfiction issue, we wanted to make sure that, at this moment of unparalleled technological advancement, we were taking the time to ask not just whether we can, but whether we should. It's high time we took stock of what we really have to lose to encroaching surveillance from our government and from corporations. As mother always said, you've got to keep your friends close and your internet service providers closer.

So, with these goals in mind, we struck out, seeking answers. We called in some of today's most incisive thinkers on privacy—lawyers, activists, journalists, whistleblowers, muckrakers—and got the folks at the Electronic Frontier Foundation on board as advisors. We learned about the work that they and others are doing to keep us safe during this time when trust is hard-earned and rarely deserved. We covered our laptop cameras with sticky notes. We cast aside Google search for its privacy-minded cousin, DuckDuckGo. (Though, sometimes we pull up Google through the DuckDuckGo search bar when we miss its cozy familiarity. Old habits are hard to kick.) We wondered more than once what government lists we'd landed ourselves on after eight months of taking calls with NSA whistleblowers and searching variations of "what percent of the dark web is drugs?" We thought about reverting to flip phones, but hell, we're no saints. It wasn't until Michiko Kakutani's *The Death of Truth* came out, though, that we realized we'd tapped into a collective awakening. Also, we'd been calling our collection *The End of Trust* since its conception years ago, and were too lazy to change it.

One unifying truth runs through the resulting collection: when our privacy is stolen, so is our right to control our own narrative. We no longer get to choose what we share of ourselves and with whom. Next to go will be our rights to speak truth to power and to express our uncensored selves, anonymously or otherwise. Most importantly, this collection reminds us that even if we feel we ourselves have "nothing to hide," even if we don't mind the DEA listening while we tell our herbalists we've recently been having more heartburn than usual, the journalists and activists we depend upon have much at stake—and so, therefore, do we. We're all part of something beyond ourselves when it comes to resisting surveillance, and the folks who are already most vulnerable disproportionally bear the consequences of our collective slide into the privacy vacuum. We need to rally together—not just because it's creepy that Taco Bell ads know what we're thinking before we do—but because privacy is a team sport, and every game counts. Every single one. We lose this or we win this together. ●

DEAR McSWEENEY'S,
Are you seeing what I'm seeing on
Instagram these days? Selfies, taken
in the feeds of store security cameras.
Sometimes they're staged in a bodega,
sometimes in a museum. They mimic
that old-school picture-in-picture
effect, which helped sell televisions in
the 1980s and '90s. A person capturing
themselves as captured by a closed-
circuit television, an image frozen in
repetition, like in a funhouse mirror.

At first glance, it might appear
to be a purely aesthetic choice, an
analog trick, as people grow bored
of the prepackaged filters on photo-
sharing apps like Snapchat and Snow.
Social media is designed to inspire
our participation, to encourage us to
reveal information about ourselves,
despite ourselves. But can I tell you
a secret? I think something deeply
profound is happening. Something
monumental. Something that entirely
reimagines selfhood. These images
are purposely reclaiming the state's
definitions of us—a radical act of play,
in the lineage of filmmaker Hito Stey-
erl's keen observation that politics
and personhood can be articulated
through imagery, ultimately pointing
toward utopian possibilities.

Did I mention that most of these selfies are of black people? Well, they are,
and usually black womxn specifically.

These images subvert the rules that
govern what it means to inhabit the
black female body, especially online.
Generally speaking, avatars have long
functioned as commentary on identity.
In *Embodied Avatars*, Uri McMillan
wrote that we are "canvases of representation" and that, in mutating our
likenesses, we are wrenching open a
new consciousness with these "brave
performances of alterity." Avatars have
the power to chart the course into a
new modality of blackness.

Security cameras have a long and
complicated relationship to blackness,
rendering it in high-contrast hypervis-
ibility, implying that the black body is
born deserving to be watched. In her
book *Dark Matters*, Simone Browne,
a professor of African Studies at the
University of Texas at Austin, links
surveillance in the West to the transat-
lantic slave trade, citing it as a source
of historical trauma for black bodies.
Quiet resistance took the shape of
something she calls "dark sousveil-
lance," a defiance that often appears as
cooperation but is—in reality—masked
retaliation against the power structures that seek to harm black folks.
These selfies are acts of resistance,
little squares of coyness looking at
the state, volunteered to a third-party
platform. A quiet and covert protest
that reverberates on a loop. "There is

liberatory knowledge in knowing how to subvert [anti-black spaces], resist or mainly survive within it and live still," Browne said in an interview in 2016.

These images are slightly jarring, like a wand I once saw used at a fetish party. We know this shock factor is the currency of today's social media climate. Susan Sontag once wrote that the onslaught of imagery saturating our daily lives "is part of the normality of a culture in which shock has become a leading stimulus of consumption and source of value." In other words, standing out requires extremism, even if it means coming dangerously close to danger itself.

There's something else about these images. They are quiet. They are articulating an independence, an imagination, but they are not loud. Their faces, upturned into the camera, unlike someone trying to avoid detection, are defiant and proud—they invite mystery. The academic Kevin Quashie would say that these women are subjects whose "consciousness is not only shaped by struggle but also revelry, possibility, the wildness of the inner life." Our world fetishizes public performance. Big displays cannot be ignored, or even avoided. But the opposite—quiet presence—can be "a surrender, a giving into, a falling into self."

We are always in modes of capture. There is no escape. Surveillance is constant, and the modern experience of blackness oscillates between the extremes of invisibility and visibility. These selfies offer an awareness of that liminal space—at once a middle finger to the state and an offering to the deities of change. I love them.

In love and resilience,

JENNA WORTHAM
BROOKLYN, NY

DEAR McSWEENEY'S,
This one time, I found a KFC logo on Google Earth. It was 2008 and I was trying to look at Area 51 in Nevada (just like probably a million other people). If not paranormal, it was definitely abnormal, this aproned colonel smiling into outer space. Most of all, I was unsettled by the logo's sharpness compared to the dusty desert surrounding it. It almost looked like an icon, but when I tried to click on it, nothing happened.

I googled "KFC google earth" and found a press release. Apparently, years earlier, KFC had unveiled what it claimed was the world's first "astrovertisement"—a giant logo made of sixty-five thousand red, white, and black plastic tiles. Strangely, the press release (now gone) was written in the UK, so they kept comparing its size to

British things like Stonehenge and Big Ben. "The largest stones in Stonehenge are the Sarsen stones which measure 8ft wide by 25ft long," they wrote. "Based on this measurement, you would be able to fit 435 stones into the KFC logo."

Fifty people took three weeks to arrange the tiles. If you search "KFC face from space" on YouTube you will find a very satisfying time-lapse video of the Colonel being assembled row by row, with some cars and a few portable toilets off to the side. They start from the bottom of the logo, so it looks like an upside-down version of a JPEG loading in 1995. At the end, the toilets disappear and all of the cars drive away.

The video convinced me that, indeed, the Colonel's image was affixed not to a picture of the Earth, but to the Earth itself. Well, sort of. By the time I saw it, it was actually gone from physical reality, having been removed six months after it was installed. During those six months, someone in Rachel, Nevada, might have walked out of the Area 51-themed Little A'Le'Inn and seen a meaningless sea of plastic tiles. Then some people (the same people?) came and took the plastic tiles away. Yet here I was, looking at them. I was reminded of the feeling I have when I look at the stars, knowing that what I'm seeing is out of date.

Honestly, though, this slippage between past and present was something I was used to as a Google Earth addict. It seems a bit smoother now, but back then Google Earth was made up of patches taken from noticeably different sources at different times. Hence the places where Interstate 80 would pass abruptly from a dry summer to the whitest of winters, lakes that were half empty and half full, and cities whose shadows fell in two different directions. I'd already begun to ascribe to these places some kind of reality of their own. Having initially gone to Google Earth for a picture of my physical world, I found instead another world—a patchy, mysterious, time-warping one with partial seasons and logos that linger indefinitely. The time in which I flew over the patchwork mountains was its own time, outside of the time of the map.

So it didn't matter to me that the Colonel wasn't really there anymore. And it certainly didn't matter to KFC! The logo was built to be seen from space, to be registered by a satellite. You could say that the minute it was launched into visibility by Google Earth, its persistence on the ground became merely incidental.

Then, years after disappearing from the ground, the logo disappeared from Google Earth. I first noticed this in 2014. At the intersection of Old Mill Road and Groom Road, there were now just some mysterious and unreadable squiggles. Why did it make me sad? I don't even like KFC–I'm pescatarian. Maybe it was just the reminder that these days a lot of things disappear not once but twice.

I'm looking at the spot now, and I notice that later in 2014, someone named Junxiao Shi made a photo-sphere (a 360-degree panorama) in front of the Little A'Le'Inn. I pick up my Pegman, dangle him over the photosphere's blue circle, and drop into the map. In the blinding sunlight, an old man in khakis is walking toward the motel. In map time, he will always be walking–walking and walking, and never getting there. I scroll 180 degrees to face the former site of the Colonel. I zoom in. I see nothing but a few small signs and, in the distance, purple mountains dissolving into pixels. I wonder if they still look that way.

Ever yours,

JENNY ODELL
OAKLAND, CA

DEAR McSWEENEY'S,
Screeeee. That there is the sound of the old man dragging out the soapbox, so if you'd rather not hear an old-timer ramble, now's a good time to turn the page, change the channel, plug yer ears, etc. Okay, you've been warned. Question: what happened to the quiet places, the quiet spaces, where a person sat and discovered who they were? When I was a young man in line at the pharmacy you can bet I wished I had a miniature TV in my hand. I just wanted to watch *The Partridge Family* or some other pablum, didn't even dare dream that I could talk to my best friend on the thing, sending cartoons of smiling faces and ice cream cones. In my wildest fantasies I wouldn't have thought that I could just tap the damn thing and con-jure up a lady's shaking rump, a recipe for banana scones, a teenaged peer weighing in on the happenings of the day with all kinds of motion graphics circulating 'round his face.

But that's what you got now, and just about everybody's got one. I don't. I mean, I don't even have a landline right now. If you wanna get in touch with my ass you gotta know when I'm at Chilly Willy's, call there, and hope that the mean ol' bartender's willing to hand the receiver over. But I'm sure if I could afford one, if I had *that* kind of lifestyle, I'd have one of those future-phones, too. I mean, kinda seems you got to. But do you *got* to look at the thing all damn day? I mean, here's an example of some of the nonsense I've observed just in the last week alone. First off, just so you can get into the groove with my terminology, my good buddy Jack has got a name for modern people who are on their phones all day, and particularly in dark old Chilly Willy's. He calls 'em ghosts. First time he did I was like, "What? Why ghosts? 'Cause they're dead to you 'cause you don't agree with that degree of technological engagement?" And Jack was like, "Nope. They're ghosts because they just sit there in the corner not interacting with anyone. And their faces glow."

Probably goes without saying, but I liked that. I liked it a lot. So that's what I call the phone-maniacs now, too. Anyways, I'm at Chilly Willy's, nursing a warm one, when I see a cute couple out on a date. Now, the both of them are ghosts. They're glancing at each other from time to time, but mostly they're locked onto their phones. It's not like I've never seen this before, it's not like I'm a time traveler, but for some reason tonight it gets me thinking, What the hell are they looking at? I mean, what is so damn interesting that they are forgoing the pleasures of the company of the opposite sex for it?

So I grab a pool cue, just kinda amble over there to get a glance at what the guy is looking at. And you won't believe it but he is looking at a picture of HER! Of the same girl he's sitting with! I'm thinking, what in the hell? Anyways, figurin' it'd take a lifetime to puzzle out that one, I walk over and take a quick glance at what the girl's got going on on her phone. First I think my eyes are fooling me, but this gal is playing a video game where you wipe a woodchuck's butt to make it giggle. Seriously, that's it. That's the sum of it! She's got the sound off, at least she knows enough to have some modicum of shame about it, but I can see the little cartoon bastard squinting his eyes and raising his paw to his bucktoothed mouth as he chuckles, as the piece of toilet paper goes sailing up and down between his butt cheeks. Up and down and up and down and... *blech*. Shoulda heeded the old saying about curiosity killing the cat, 'cause I'll admit, it got to me. Spent the rest of the evening quite blue. I mean, what kind of effect is this constant flow of audiovisual bullshit having on my fellow man? Seriously. It troubles me.

I know we had newspapers and paperbacks and the dumb game solitaire to suck up spare time before these phones, but think about how many times you used to see lonely folks just staring out the window, or at the carpeting, or into their clasped hands. Think about how often *you* did that. And I just keep thinking, what happens now that we've eliminated the quiet spaces where a person thought hard about things, where they discovered the depths of themselves, like it or not?

I keep thinking back to this one time when I was a teenager and trying to get my first kiss. To say I was lagging behind my peers is to put it lightly. I was fifteen and still could count on one hand the times a girl had let me rest my leg against hers. There were reasons for this I don't want to get into right now, but point is a very pretty girl named Amanda had finally agreed to meet me down by the creek. The creek had a reputation, as did Amanda, so I was sure I was about to get a kiss. Went down there at sunset, left alone 'round midnight. I think it goes without saying that Amanda never showed. And the whole time I was waiting for her I didn't get to send out a flurry of text messages about it, or watch celebs waggle their rumps, or open an application and peruse a million other girls, "swiping" them this way and that based on the whims of my attractions. No, I just had to sit there and stare at the water and wonder why she wasn't coming. First I wondered if something had happened to her, then I realized that was unlikely and had to reckon with the fact that she'd decided to stand me up. Why? I asked myself. Why

would she do that to me? And the question forced me to look at some of the behaviors I was exhibiting at the time. And while I would have given anything for a kiss in that moment, I think the lack thereof forced me to confront some hard truths, and to grow up a little bit. So again, I can't help but wonder, what happens when those spaces are gone? What happens to a people constantly pacified? You end up with a bunch of goddamned babies. You end up with ghosts.

Yours,

CARSON MELL
LOS PALACIOS, CA

DEAR McSWEENEY'S,
He was in his mid-forties, but like most men in the early 1800s he looked much older. His hair was white and balding; he had formidable sideburns. His cleft chin was small and pushed out from folds of skin. This was all swaddled with a white gauzy scarf. At least, this is how the portrait artist painted him. His look was serious but not stern, old and important and neutral. He could be on a coin or hanging in the lobby of a bank.

His name was Franz Joseph Gall. He was a Viennese scientist, credited with inventing phrenology–the study of the shape and size of the head as an indication of character and mental abilities—a discovery which he claimed to have made as a schoolboy. This is how he put it: looking around the classroom, he noticed that the students who were best able to memorize passages had larger eyes and prominent foreheads, and in that moment he decided that the truth of human cognitive capacity must lie there, behind the eyes. He would spend the rest of his life developing a "perfect knowledge of human nature."

The desire for *a perfect knowledge of human nature* could also reasonably describe our modern infatuation with using computational technology to solve age-old questions about human psychology. The most popular books about data science include titles like: *Everybody Lies: Big Data, New Data, and What the Internet Can Tell Us About Who We Really Are* and *The Most Human Human: What Artificial Intelligence Teaches Us About Being Alive.* They pledge entirely new ways of understanding the human condition. They imply that we need someone to corral this data (it's everywhere!), and tell us what it says about us. Give us *a perfect knowledge of human nature.*

People made connections between the shape of heads and character long before Gall first assessed his peers' intelligence. Take the Buddha, for

instance; some statues of him show a large cranial lump, which represents knowledge and attained wisdom. Or Aristotle, who wrote in *Historia Animalium*, "When men have large foreheads, they are slow to move; when they have small ones, they are fickle; when they have broad ones, they are apt to be distraught; when they have foreheads rounded or bulging out, they are quick-tempered." Seneca, Cicero, and Shakespeare were also known to relate personality to body and head shape.

Phrenology became wildly popular during the nineteenth century, at the height of Romanticism, which reimagined man as "natural," took importance away from inherited knowledge, and focused instead on character and emotion. This began a new process of making the inner, invisible things visible. And this process, where one's physicality became the pathway to the soul's mysteries, required new tools. For the early phrenologist, the tools were bare fingertips, rubbed over a patient's head in order to distinguish elevations or indentations, which corresponded to approximately thirty-seven "organs" (Gall's term) and indicated personality traits like marvelousness, destructiveness, and combativeness. This information was tabulated and interpreted; phrenologists created complex narratives from the data they collected. They made predictions for people's futures and prescriptions to keep their less desirable traits in check.

The church called Gall a heretic, which made his ideas that much more attractive for the secular, or semi-secular—at odds with the church's strict dogma, but still willing to place their faith in the unseen. What if, instead of in church, powerful spiritual faculties existed in science? Cathy O'Neil, author of *Weapons of Math Destruction*, says that our relationship to data is similar to the way one might put faith in God. People are prevented from asking questions, even when the data ends up reinforcing discrimination and widening inequality. "I think it has a few hallmarks of worship," she said in an interview with the *Guardian*: "we turn off parts of our brain, we somehow feel like it's not our duty, not our right to question this."

In 1832 Gall's protégé, Johann Gaspar Spurzheim, brought these ideas from Europe to America, a country hungry for notions of individual betterment—a precursor to the self-help craze. Spurzheim was only in America for six months before he died, but in that time he gave hundreds of lectures at prominent institutions—Harvard, Yale, etc.—and converted thousands. He was described as one of the greatest minds by Ralph Waldo Emerson, and when he died John James Audubon sketched his remains and the *American Journal of the Medical Sciences* declared, "The prophet is gone."

The desire to develop a *perfect knowledge of human nature* isn't, on its surface, hateful; however, physicians used phrenology to say that the skulls of African people indicated that they were mentally inferior and therefore suited to be slaves. Native Americans were deemed slow, a point that Andrew Jackson used to justify his removal policies, and eugenicists in Nazi Germany used phrenological research. These tools were also used to predict criminal activity, and in the process signaled a shift in the development of a new "science" of crime—a shift which persists today.

A couple of years ago, while researching labor at the Maine State Prison in Thomaston, I visited the town's small museum. In a glass case, a long row of phrenological tools was displayed. The prison once used these tools and the data they produced as evidence, which could mean life or death, freedom or enslavement. If the small plaque had not been there, the tools might have looked like something old, unidentifiable, and broken—a pile of brass bars and rulers.

Philadelphia, where I now live, was once home to Samuel George Morton, the nineteenth-century naturalist who collected skulls from all over the world and meticulously catalogued their similarities and differences. (He and his colleagues were known to rob graves and trade the skulls of indigenous people, some of which are said to still be housed in Philadelphia's Academy of Natural Sciences.) When Morton deduced that "Caucasian" skulls had the largest cranial capacity and were therefore superior, he did so with full faith in numbers. Historians have a bad habit of treating phrenologists like eclectic dilettantes or fringe amateurs, when many of them were highly educated and respected "men of science," who did not simply misunderstand the data before them. Morton was no amateur. He was adamant about his neutrality and scientific process, qualities we still believe to be integral to real science.

Several years ago, jurisdictions across the country began using early forms of risk-assessment algorithms to predict who would commit crimes or not show up for court, among other outcomes deemed unfavorable. These tools have undergone many changes over the years. Now many use machine learning, which means that the tool teaches itself how to make better predictions: a *perfect knowledge*. When tools use machine learning, even the statisticians who created them, after

a certain point, can no longer tell you easily how data is being used to make determinations. In this case, it would be hard, if not impossible, to know if a tool was using a factor like race to tell a judge whether or not a person was "dangerous" or "not dangerous."

Proponents of risk-assessment algorithms claim that there are responsible ways to implement them. The answer doesn't have to be incarceration, but access to resources. If an algorithm tells us a person has a high risk of committing a crime, well, we should check in on that person; give them access to drug treatment, should they need it, or a counselor.

Gall also believed that there was a best practice to the science he developed. He believed that when someone committed a murder, the act was a struggle among the different "brain organs," and therefore it would be impossible to say definitively who was capable of what; even though we could not change the innate limits of our cerebral organs, we could–through education and access to opportunities–give more power to the organs of higher motives. In other words, according to Gall, each person carries the propensity for best and worst possible outcomes, and it's up to us to create an environment most conducive to the former. But, of course, this isn't how things worked.

Phrenology and risk-assessment algorithms aren't one-to-one, but there are similarities in rhetoric: both ask for a character report, both posit that there is a subject to be read and that, in the interest of public safety, that subject must be surveilled and made legible. We can note the similarities without claiming that one is inherently like the other. We can also note that we are still hungry for prediction and downright ravenous for a *perfect knowledge of human nature*, when we should be deeply skeptical of the process that turns collected data into narratives, and the unknown ways in which we might read, calculate, and assign interpretive power in the future. Using data to come to conclusions about human nature isn't new, but the age of Big Data is, and it comes with a new language. New languages bring new currency and power–but for whom?

Still asking,

CHELSEA HOGUE
PHILADELPHIA, PA

DEAR McSWEENEY'S,

"Alexa's listening to you right now," John says to me the other day. "She isn't supposed to be, but she obviously is."

(Alexa is a phallic speaker tube with a light ring, which sits on our dining room sideboard like a vase of flowers, though, in fact, in front of a vase of flowers. She is an affordable AI who looks a little like a pocket version of the monolith from *2001: A Space Odyssey*. She is supposed to simplify lots of tedious life tasks for people via voice interface, but mostly all she can do is set a timer or play a limited selection of music through her shitty speaker.)

"She's always listening?" I ask.

"They say she isn't. She's always on, but her name is her wake word, so theoretically she only begins to record what we say when she hears her name."

"Record?"

"She records all her interactions and reports to the cloud. That's how she develops. I get a transcript on the app. You'll get one too. She only records the things you say after you say her name, so it isn't technically an invasion of privacy, since you are inviting her into discourse."

"That's just what J. Edgar Hoover said to me in the bar last night," I say.

"I basically agree with you. You know I'm only doing it for art."

"That's the other thing J. Edgar Hoover said to me."

John was offered Alexa as part of an experimental launch, and so he has invited this tube into our house for artistic purposes, he claims. The artistic potential of Alexa is dubious. The potential for constant surveillance, however, is certain. I grew up during the end of the Cold War so I have been conditioned to fear surveillance, both foreign and domestic, though probably more domestic than foreign. Initially, thinking about Alexa, I would start to hear the slightly off-key piano theme to *The Conversation* and turn on all the faucets and close all my blinds and hum really loudly, which of course is pointless, since she's already inside, and can supposedly cancel out ambient noise in order to better serve us. But pretty quickly her espionage talents seemed compromised by the fact that she mishears everything. We get reports back on what we have said, or what she thinks she has heard, and what the cloud has recorded. It's all there in the cloud, inaccurate, poorly spelled, and affectless. Things come back so muddled, it seems like John and I are already speaking in an encrypted code language. The theme music of *The Conversation* dissipates and is replaced by the theme music from *The Pink Panther*. Ultimately, Alexa may be more a pataphysical machine than a tool for subterfuge and reconnaissance, and so perhaps she only makes sense for art.

Why she has been gendered female is quite beyond me given her physique, though I assume it's part of a marketing approach that plays to an internalized sexism in which all secretaries are women and/or all women are secretaries, and, of course, hapless.

Nonetheless, John and I whisper when we are within tubeshot of Alexa. I don't know why we whisper, since she often can't hear us even when I am shouting right into her tube, which I do often, in a rage, because she is incapable of doing anything. I shout at her in a rage when she can't find the Hawkwind album I want to hear through her shitty speaker, and I shout at her in a rage when she is playing neoliberal news reports despite my repeated efforts to change her settings to more radical alternative news sources, and I shout in a rage because she's misunderstood my request for the weather report and has instead added weather to my shopping list. When I am at home alone with her, I yell at her a lot. I ask her for things I know she can't possibly do, and then I yell at her when she fails. I speak to her in languages I know aren't in her settings, and then I yell at her for her monolingualism. I request information, and she misunderstands, and I yell at her. "If you can't understand anything, or do anything, or find anything, Alexa," I yell, "maybe you can just shut up and play some smooth jazz!" and she says, calmly, flatly, "Shuffling smooth jazz playlists from the cloud," and she plays some smooth jazz, and I berate her for it, even though it isn't unpleasant.

As I read through the transcripts of our conversations with Alexa, several things become clear to me. First, in voice-recorded surveillance, assessment of wrongdoing is still extremely dependent on human interpretation. Machines typically don't notice if you seem to speak in gibberish. If you want to expose someone's crimes via the tube, you will have to be able to interpret and intuit their criminal intent amidst the word salad that has been cobbled from the scraps of voice. This aspect of my research suggests that the role of humans in the rise of the machines will be that of interpreters, informants, and spies, betraying any human resistance cells, and exposing them to the machines. That much I could have guessed without Alexa. However, I've also learned that, contrary to Foucault's suppositions about the constant surveillance from the panopticon, it doesn't seem to lead me to self-policing where Alexa is concerned. I will shout at her and berate her, knowing full well that my misbehavior toward her is being documented. This is the irrational disregard that will likely get some of us slaughtered by the machines before others.

Even when he's at the office, John gets the transcripts on his app telling him what Alexa thinks I am saying to her. He can intuit the affect in my voice even from the muddled code. If I scream at Alexa over my second coffee, John will check in by midday to police my abuse of her. That's how well he knows me. Just as, when I read transcripts of John's conversations with Alexa, I can tell that he is asking her about popular culture items which he feels are youth-specific and threatening in their mystery. He would rather ask Alexa than admit to me that he's out of the loop.

"Stop being beastly to her," John says when he checks in.

"Stop spying on me with your surrogate phallus," I yell.

"Since when do you listen to Hawkwind?" he asks.

"Since when do you 'Hit the Quan'?" I yell.

Our intelligence is organic! We feel savvy in our knowing and interpreting and understanding of our own voices as they move through the circuits—an intelligence, or willful lack thereof, that will almost certainly doom us when the machines rise.

Be seeing you!

JOANNA HOWARD
PROVIDENCE, RI

t didn't have to be like this. The Electronic Frontier Foundation was founded by John Perry Barlow, Mitch Kapor, and John Gilmore on the recognition that technical architecture is politics—that how we build our tools will determine our rights. Technology itself has no political positions. Our digital world can be fair or unfair, empowering or disempowering, utopian or dystopian depending on the choices we make along the way.

Yet for the last decade we've seen technology tend toward the unfair, the disempowering, the dystopian. We've seen governments and companies take negative advantage of their positions in building and running the networks, architectures, and tools that the rest of us rely on. They treat us as unimportant serfs in their mass spying systems, fodder for machine learning algorithms—and treat our world like a cybersecurity battleground where our private lives are mere collateral

EFF has worked to make sure that your rights go with you when you go online. We use pre-digital ideas—like those in the Bill of Rights and the Universal Declaration of Human Rights—to make sure that the digital world bends toward empowerment and fairness. We use the courts, try to reach the lawmakers, and push the companies in a positive direction. We've represented makers, archivists, coders, and culture jammers, working to ensure that the smaller engines of innovation and knowledge have the necessary space. We've grown a tremendous amount, over forty thousand members strong at last count, but sometimes it feels like the underlying ground has grown so rocky and hard that, while we can sometimes stop bad ideas, good ones never get to grow.

Law enforcement and intelligence agencies have marshaled the politics of fear and the power of secrecy. They have too often convinced our courts and legislators to sign off on—or refuse to even look at—public and private technologies that strip us of our privacy and our ability to organize free of prying eyes. Technology companies have embraced the surveillance-business model, reducing much of the digital world to a small number of centralized, controlled, surveilled spaces. Together, governments and corporate powers have conspired to monitor our private communications. Even when they don't have access to the conversations themselves, they can learn whom we talk to, when, and how often. They have ensured that we remain mainly consumers—sharing rumors, stoking outrage, and clicking away at ads that follow us from page to page as we look for things to buy—rather than organizing and engaging in real self-governance. Messages like "privacy is dead, get over it" and "free speech is just a cover for hate and harassment" serve as a kind of autoimmune disease attacking our most cherished rights.

But we cannot retreat into privacy nihilism. Major scandals do periodically outrage us—like Mr. Snowden's leaks, Cambridge Analytica's misuse of Facebook-enabled personal data, and the toxic brew served to us in the 2016 election by algorithms and dark social engineering of our social networks. We must use those moments of outrage to build political power, even as forces work to steer us back into hopelessness or to retreat into our ideological bunkers.

From where we sit we see signs of hope. Many, especially younger folks, are picking up the charge. Some are learning to build their own tools or contribute to ongoing projects building a better architecture for all of us. Others are holding events like cryptoparties to teach each other about security. Too much is at stake especially for those who don't live in the wealthier parts of white America, to let our technological world become merely a shopping mall where we buy things and exchange half-truths. Those who find themselves the butt of attempts to create a scapegoated "other" are learning to use those same tools to fight back. We see it in the kids organizing online against gun violence and in the increasing public documentation of police abuses against Black people. People from all walks of life are using—and teaching others to use—the cryptographic tools that create real security for organizing communities. An increasing number of people recognize that tremendous political and financial power has stoked hatred and outrage. With that recognition come shifting winds.

Standing up for a better technological world is harder, but also more important, in a society where trust has been eroded, battered, and devalued. As hill farmers say, any fool can farm flat land. Well, any fool could push for a better world if it were right at our finger tips. But we need to be the hill farmers of the digital world.

The task is not easy, but the cause is not lost. Trust can be rebuilt. We must be willing to build and invest in an internet that protects privacy, provides a voice for the voiceless, and allows us to change our futures through activism and assembly. The digital revolution lets people meet and learn from each other, regardless of physical borders and boundaries. It lets us learn anything we want to learn. We can still harness that power.

EFF has been here all along. We've continued to use the tools of law, activism, and technology to build a more trustworthy future, even as we decry how bad things have actually become. We push court cases that serve users. This includes helping eliminate the Third Party Doctrine, which says you surrender your constitutional privacy rights when you let third parties hold your data. The U.S. Supreme Court even gave EFF a shout-out in June 2018, when it took a big bite out of that outdated doctrine. We also work to ensure that your Fourth and First Amendments rights apply when you carry devices over the border. We draw attention to and push back against the growing list of military-built surveillance tools being used domestically—from license plate readers to IMSI catchers to facial recognition. We represent the coders who are building better tools and identifying insecurities in our current ones. We also build technologies to encrypt the web, like Let's Encrypt, HTTPS Everywhere, and STARTTLS Everywhere. We created Privacy Badger, a tool that helps users regain control of their web browsing by blocking those pesky trackers that follow you from site to site. We push companies to stand with their users, using projects like Who Has Your Back which rates technology companies on how well they protect their users' data from the government. And we are growing, as more people, including less technical people, recognize that they need to commit to fostering a technological world that we actually want to live in.

But enough about us. We have been delighted to assist with this collection because we know that in order to grow toward the light we need more than just smart lawyers, committed activists, and clever technologists. We need writers and artists and readers like you. We need everyone.

The first step is to recognize the problem, and the values at stake. Though it's sometimes dark work, this is imperative. It includes telling people how bad it has become and clearly recognizing how surveillance disproportionately impacts already marginalized people. We are happy that this issue features some of the most insightful voices on that point, including Alvaro Bedoya and Malkia Cyril. Privacy nihilism so often comes from a place of privilege; far too many of us simply aren't capable of tuning these issues out.

Several authors—including Ethan Zuckerman, Douglas Rushkoff, Ben Wizner, Edward Snowden, and Gabriella Coleman—reflect themes of decentralization and reliance on nontraditional power structures. Other authors talk about using technology for ends that serve the people, building literacy and organizing, like Thenmozhi Soundararajan, Camille Fassett, and Soraya Okuda.

We may not agree with every point made in this collection, but exploring these themes is important. We hope you'll join us in the work of envisioning and building a trustworthy future. We don't say it will be easy. But any fool can farm flat land. ●

Everything Happens So Much

Sara Wachter-Boettcher ➔

"**G**oogle in Boston is interested in hosting you for an event," my publicist emailed me one day last September. No pay—not even travel expenses to come up from Philly. But they wanted to record my presentation for their Talks at Google series, they said, and they'd buy some books to hand out.

Normally, I'd decline this sort of thing: a tech giant that's vying to become the first company worth a trillion dollars wants my time and expertise for free? *lol nope*, I'd type to a friend, throwing in the eye-roll emoji for good measure. But the book, *Technically Wrong*, my first mainstream title and something I was deeply nervous about, was launching that month. And given its topic—a look at how a toxic culture within the tech industry leads to products with bias, exclusion, and all sorts of unethical behavior baked right into them—it seemed like something I shouldn't pass up. "I spoke about this at Google" adds a layer of credibility, and the video a bit of visibility.

So here I am, trudging across Cambridge on a windy October morning, presentation clicker in my hand, gnawing anxiety in my stomach. It's only been two months since the press picked up the Google Memo—former engineer James Damore's notorious ten-page screed arguing that women are simply biologically less capable of programming than men and demanding that Google end its diversity efforts and stop training staff on skills like empathy. I assume they invited me as part of some sort of PR crisis-recovery plan. But I also know Damore has plenty of fans inside Google, and I'm not sure what kind of reception I'll receive on the ground once I get there.

Meandering through the cafeteria—sorry, *campus café*—before my talk doesn't help things. All around me are groups of guys in polo shirts and khakis. Groups of guys in T-shirts and hoodies. Groups of guys in flannel shirts and Warby Parker frames. I can't say I'm surprised, exactly—the Boston office is mostly engineers and Google's technical workforce is still, after years of touting its diverse hiring efforts, 80 percent male. But I'm on edge, hyper-aware of the distance between me and them.

I slide into a seat at a table, a bowl of Google-subsidized organic kale and chicken in hand, and think about the people around me. Do they realize they move in packs? Have they ever noticed how many men surround them, day after day—and, correspondingly, how few women? Will any of them even show up for my talk?

I get onstage anyway, and the talk goes fine: I talk about photo recognition software that can't recognize black people, app notifications that leave out queer people, and a tech industry so hell-bent on "delight" it hasn't bothered to ask whether what it's doing is ethical. The attendees ask smart questions. A group lingers afterward to chat. I step back out into the city relieved.

A month later, Google lets me know that the video has just gone live. I click the YouTube link.

> This is exactly the kind of hysterical, sanctimonious female that got Prohibition into law.

> This woman is mentally ill... She needs medical care, not a microphone at Google.

> That "lady" needs to not wear those pants.
> It's offensive to my eyesight.

I close the tab, smart enough to know that nothing's gained by reading strangers debate whether you're *thicc* or just plain fat. And then I cry anyway.

It isn't the vitriol that breaks me. I knew that would happen when I started writing my book. In fact, I thought it might well be worse—I'd prepared my accounts, warned my husband, asked my publisher how they'd help if I were targeted (they didn't seem to understand the question). It's the speed of it: before I could even share the link with my friends, there were already dozens of comments like this. It had been posted on some forum or other, and a small army in the ground war against women had been sent to put me in my place. Standard practice, to be honest.

I should know. I wrote thousands of words outlining the ways tech products were designed to allow, or sometimes even encourage, abuse. We can see it at Twitter, where—more than half a decade after women started reporting systematic harassment on the platform, four years after Gamergate made Zoe Quinn's life a living hell, and two years since Milo Yiannopoulos sent a legion of trolls to attack Leslie Jones for daring to be a black woman in a *Ghostbusters* movie—executives still have no idea how to curb abuse on their platform, a place where Pepe the Frog avatars and Nazi apologists run amok. We can see it on Reddit, where subreddits teeming with virulent racism are expected to be handled by unpaid moderators—"Make the users do the hard part," as former general manager Erik Martin put it. And we can see it, as I did, on YouTube, where misogynists (and the bots working for them) flock en masse to content about social justice issues, aiming to immediately downvote the videos themselves while upvoting the vilest comments on those videos. The game is simple: if you can quickly make the content appear unpopular, YouTube will show it to fewer people. And those who *are* shown the video won't just see the talk; they'll also see those top-rated comments.

I'm not just aware of this system. I've painstakingly mapped it out, looking at the way young, financially comfortable white guys come up with ideas that work great for them—without ever noticing just how badly they'll fail for people who are marginalized or vulnerable to abuse. And then, once they *do* notice, they flail about for months or even years without really getting anything done.

The insults themselves aren't even surprising. They read like a page from a sexism primer: I'm ugly, I'm irrational, I'm a humorless nag. It's all so obvious. I've had thirty-five years to learn the rules, after all. I know that to be taken seriously, I should erase my feelings, thicken my skin, avoid being *dramatic*—or expect to have my intelligence and mental health questioned. I know that if I want to avoid ridicule, I should hate my body, too—I should cover my arms, rein in my thighs, wear more black, be more invisible. I had betrayed all those teachings. I'd been confident. I'd talked about feelings. I'd gone onstage in *purple pants*, for chrissakes. And the trolls knew exactly how to punish me for my transgressions. They'd learned the same rules I had.

None of it should feel personal, but of course it does. It always does. But it's also, somehow, validating. Like waking up to a raging head cold after three days of questioning whether or not that tickle in your throat was real. I wasn't imagining things. The sickness had been festering this whole time.

I didn't trust tech companies much before I started writing this book, and I trust them even less now. The big ones care about pleasing shareholders. The small ones care about pleasing venture capitalists. None of them care about you.

But what I didn't anticipate is how technology would also erode my trust in myself.

Ages ago, in those sweet years before fake news and Russian election hacking, before Gamergate, before random men showed up in my mentions every day to explain my own work to me, Twitter was my lifeline. It was a way to find and connect with peers in an industry–content strategy and user experience design–that was only just emergent, and that I was only just beginning to claim as my own.

It was also a way to bring together my personal and my professional sides, teasing out a space for myself that felt smart and authentic. I could be funny. I could be earnest. I could share an article I'd written about metadata on Tuesday morning, and then send a series of half-drunk tweets about a TV show that night. I felt seen. I felt understood.

Ten years later, I hardly recognize that person. One day, I type drafts and delete them, watching the stream go by without me. The next, I share praise for my book, or I link to the new episode of my podcast, or I retweet the latest article I've written–and I feel ashamed of my self-promotion. I vacillate between a need to share my voice–to *use my platform*, as they say–and a growing desire to hide.

I don't just feel seen anymore. I feel surveilled. Judged. Anxious about what it all means. I calculate: am I making myself a target? Is this the tweet, is this the opinion, that will finally bring on a wave of red-pill trolls and angry white supremacists so big it bowls me over for good? Is feeling surveilled the price I have to pay for being ambitious, for wanting to create and critique and participate in the world? What does it say about me that I'm willing to pay it?

And then, if we're being honest here, I also think something much darker: why am I not getting more abuse? Is my work too ignorable? Are my opinions too safe? Shouldn't I be more controversial by now?

I berate myself. What sort of monster feels *jealous* of people who are being harassed?

This one, it turns out.

Back in 2011, I became infatuated with @horse_ebooks. Purportedly a bot run by a Russian spammer, the account regularly tweeted absurd text snippets: "Unfortunately, as you probably already know, people." "Get ready to fly helicopters." And my personal favorite: "Everything happens so much." The tweets were mesmerizing, inexplicably hilarious, and wildly popular. They were Weird Twitter at its finest.

They also weren't generated by a bot–or at least not for long. Apparently the Russian spammer sold the account to a Buzzfeed employee sometime the same year I'd discovered it, and he'd been the one writing the tweets ever since. Yet, years later, that line still rattles around in my brain. Everything happens so much. I've even found myself unconsciously whispering it out loud as I scroll through my feed, overwhelmed by breaking news and conversations and jokes and trolls and cats and everything else.

Everything happens so much. That's the beauty, but it's also the problem. It's not that technology broke my trust–at least not at first. But it broke my context: I don't know where I am. I don't know whether I'm at work or at play, whether I'm watching the news or chatting with a friend. This used to feel freeing: I didn't have to choose. I could simply *exist*, floating in a mix-and-match universe of my own design. But left unchecked for so long–by shortsighted tech companies, and by my own petty desires–that lack of context bred something sinister: a place where everyone's motives are suspect. I don't know who's watching me, or when they're coming for me. But I do know they're there: the James Damore fanboys, the YouTube troll armies, the Twitter Nazis, the casual misogynists itching to play devil's advocate.

For now, at least, so am I. I'm just still not quite sure why. ●

A COMPENDIUM OF LAW ENFORCEMENT SURVEILLANCE TOOLS

By Edward F. Loomis

● Facial Recognition Systems

○ GPS Trackers

○ License Plate Readers

○ Drones

○ Body Cameras

○ Cell Tower Simulators

○ Parallel Construction

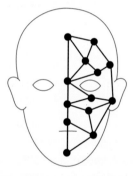

With public attention focused on the federal government's warrantless surveillance, the state and local use of surveillance methods has largely been ignored by many citizens. Following 9/11 and the subsequent creation of the Department of Homeland Security (DHS), many military-grade surveillance technologies have migrated through DHS grants into the hands of state and local police departments for the supposed purpose of thwarting terrorist acts. These high-tech tools are in everyday use by officers across the country, capturing enormous amounts of data on unaware citizens who come within their range. Woven throughout this issue are a few of those tools.

Mug shots of suspected criminals have proven useful in solving crimes since their use by the Pinkerton National Detective Agency in 1870. With the development of digital computers, experiments began that automated identity recognition through distinguishing facial features. In the years since 9/11, automated facial recognition systems greatly matured to meet the needs of U.S. troops to monitor local populations in Afghanistan and Iraq. The technology was soon adapted by the law enforcement community to monitor popular events for security purposes. With the rapid development of digital cameras, cell phone cameras, cloud computing, social networks, and public posting of tagged images, a database of millions of personal facial images can easily be created by applying available surveillance methods.

Facial recognition systems can examine the distinct features of

▷ | **A Compendium of Law Enforcement Surveillance Tools**

● Facial Recognition Systems
○ GPS Trackers
○ License Plate Readers
○ Drones
○ Body Cameras
○ Cell Tower Simulators
○ Parallel Construction

`MCS54` 0040

each individual's face in photos, video, or a real-time camera feed and match it against those in a facial database. Some systems score candidate matches and rank them in order of relative confirmation probability. However, accuracy of these automated systems varies widely depending on the race and gender of the subject. In tests conducted by a researcher at the MIT Media Lab, identifications of white males were demonstrated to be accurate 99 percent of the time, whereas the failure rate on dark-skinned females approached 35 percent.

As arrested subjects are photographed by local, state, and federal police, their mug shots are stored in that jurisdiction's database for use in future automated criminal searches. Where cooperative sharing agreements exist, such as in the San Diego County region, where a consortium of eighty-two local, state, and federal law enforcement agencies shares access to the Automated Regional Justice Information System (ARJIS), photographs may be queried by member jurisdictions to check traffic-stop subjects for outstanding warrants or to support criminal investigations.

Beginning in 2007, ARJIS received Department of Justice grant funding to pursue development of a mobile, regionally shared facial recognition prototype named the Tactical Identification System (TACIDS). Since 2013, the San Diego County Sheriff's Department has been using hand-held tablets that connect to the TACIDS facial database. Without asking an arrest subject his or her name, an officer can capture the suspect's image with a tablet or smartphone and upload the photo to the TACIDS database. The system returns positive matches from its over 1.3 million booking photos of prior arrestees for a virtual lineup.

One can imagine the privacy implications of a massive virtual lineup database supporting an automated facial recognition system. According to data compiled in 2016 by the Georgetown Law Center on Privacy and Technology, at least twenty-six states "allow law enforcement to run or request searches against their databases of driver's license and ID photos." Additionally, sixteen states permit FBI use of facial recognition technology to "compare the faces of suspected criminals with their driver's license and ID photos."

Most troubling is the fact that no state was found to have laws that comprehensively regulate the use of facial recognition systems. The study described that the Pinellas County Sheriff's Office "runs 8,000 monthly searches on the faces of seven million Florida drivers—without requiring that officers have even a reasonable suspicion before running a search." The study found that of fifty-two agencies reporting the use of facial recognition systems, only the Ohio Bureau of Criminal Investigation had policies prohibiting its officers from using them to track persons engaging in protected free-speech activities. Of those fifty-two agencies, only four had publicly available policies on the use of facial recognition systems, and only nine claimed to log and audit officer searches for improper use.

For those who have personally been mistaken for someone else by a stranger, the fact that facial recognition software might erroneously select us from a growing national lineup database can be very unsettling. When South Wales police activated a facial recognition system against 170,000 attendees at a soccer match between Real Madrid and Juventus in Cardiff in June 2017, the cameras identified 2,470 people as criminals. Of

▽ **DATA**

1/2

One in two American adults (over 117 million) is in a face recognition network run by law enforcement.

30%

MIT testing of three commercially-released facial recognition systems found a 30% lower success rate for "darker females" than "lighter males."

Source: Center on Privacy & Technology at Georgetown Law; MIT Media Lab.

the potential criminals identified, 2,297 were wrongly labeled by the facial recognition software—a 92 percent false positive rate.

In FBI testimony presented before the House Committee on Oversight and Government Reform in March 2017 on the FBI's use of facial recognition technology, an admission was made that the technology is used only as an investigative lead, not as a means of positive identification. The reason for that became obvious when the facial recognition system's accuracy was later cited as having been tested and verified only to return "the correct candidate a minimum of 85 percent of the time within the top 50 candidates." ●

Peak Denial

Cory Doctorow →

On April 22, 1990, I was one of the tens of thousands of people who packed into Toronto's Nathan Phillips Square, in front of City Hall, to commemorate Earth Day, the first one in twenty years. I was eighteen years old, and as far back as I could remember, I'd had an ambient awareness that there was an environmental crisis in the offing–from the PCBs in the water to the hole in the ozone layer. But as the day wore on, as the speakers took to the stage, as the crowd buzzed and surged, I realized that this was *serious*. It didn't matter how big and hopeless the cause was; the fragile ecology of the only known planet capable of sustaining human life in the *universe* was at risk of permanent and irreversible collapse.

That's when I became a climate bore. For years after, I tried to convince the people around me that we were hurtling toward a crisis that would adversely affect them and everyone they loved: surging seas, killing winds, waves of disease, and water and migration crises loomed in our future, and by the time they were so manifest as to be undeniable, it might be too late.

I failed. We all failed. When we mustered enough political pressure to provoke international effort, it fizzled: Kyoto, Copenhagen, Paris. All unambitious, nonbinding, insufficient, and, of course, blown past by almost every signatory.

But a funny thing happened on the way to the climate apocalypse: the job changed. Twenty-eight years ago, most people I had the Climate Talk with assumed that I was exaggerating the danger. Now, most people I talk with accept the danger–they just don't think we can do anything about it.

That's because somewhere along the way, we crossed the point of "peak denial"–the moment at which the number of people who just don't think climate change is that big of a deal started its inevitable decline. Some combination of science communication (movies like *An Inconvenient Truth*) and undeniable consequences (floods, hurricanes, droughts, blizzards, and other extreme weather events) convinced a critical mass of people that the problem is urgent.

The consequences are only getting more undeniable as the science becomes clearer, scientific literacy becomes more widespread, and the communicators sharpen their explanations. It's hard to imagine how the number of people who acknowledge the reality of climate change could ever significantly decline. From now on, the number of people whose direct experience makes them climate-change believers will only grow.

This presents a new tactical challenge for activists. It's no longer so important to convince people that climate change is real—now we have to convince them that we can and should do something about it. Rather than arguing about problems, now we're arguing about solutions, and specifically whether solutions even exist. Before it was a battle between truth and denial; now it's a battle between hope and nihilism.

It's not just climate change. Many of us went through an individual journey similar to this one with cigarettes: rationalizing a dangerous and antisocial habit by leaning on the doubt expensively sown by tobacco companies. After all, tobacco, like greenhouse gases, does not manifest its consequences right away—the cause-and-effect relationship is obscured by the decades intervening between the first puff and the first tumor. And just as with climate change, by the time the symptoms are unignorable, it's tempting to conclude it's too late to quit: "Might as well enjoy the time we have left."

Which brings me to privacy.

Breaches of our privacy share many of the same regrettable characteristics as climate change and cigarettes: your first unwise disclosure and your first privacy breach are likely separated by a lot of time and space. It's hard to explain how any one disclosure will harm you, but it's a near-certainty that after a long run of unwise disclosures one of them will come back to bite you, and maybe very hard indeed. Disclosures are cumulative: the seemingly innocent location-tagged photo you post today is merged with another database tomorrow, or next year, and links you to a place where a doctor was writing a methadone prescription or a controversial political figure was present.

Getting people to overshare is big business, and the companies that benefit from public doubt about the risks of aggregating personal information have funded expensive PR campaigns to convince people that the risks are overstated and the harms inevitable and bearable.

But the fact is that collecting, storing, and analyzing as much personal information about the public as possible is dangerous business. The potential for harm to the subjects of this commercial surveillance is real, and grows with each passing day as the database silos get bigger.

People who've had their personal information collected face a long list of risks: blackmail, stalking, government or police overreach, workplace retaliation, and identity theft are at the top of that list, but it continues on and on (for example, some breach victims had their names forged on letters to the FCC opposing net neutrality).

What's more, the harms of privacy breaches don't fade over time: they worsen. Maybe your account on a breached addiction-recovery site is tied to an alias that isn't connected to your name today, but a decade later a different site where you used the same alias gets breached, and some clever person uses both breaches to re-identify that formerly anonymous alias of yours.

Privacy is a team sport. The data from my breached accounts may compromise *you*, not me. For example, your messages to me might reveal that you contemplated divorcing your spouse years before, a fact that is revealed when my data is breached, years after you'd both reconciled.

Add to that the effect on social progress. In living memory, it was a crime to be gay, to smoke marijuana, to be married to someone of another race. Those laws gave way after massive societal attitude shifts, and those shifts were the result of marginalized people being able to choose the time and manner to reveal their true selves to the people around them. One quiet conversation at a time, people who lived a double life were able to forge alliances with one another and with their friends and family. Take away people's right to choose how they reveal themselves and you take away their ability to recruit supporters to their cause. Today, someone you love has a secret they have never revealed to you. If we take away their ability to choose how they reveal themselves to the world, you may never learn that secret. Your loved one may go to their grave sorrowing without your ever knowing why.

Which is all to say that every day that goes by auto-recruits more people to the burgeoning army that understands there is a real privacy problem. These people are being recruited thanks to their visceral experiences of privacy breaches, and they're angry and hurt.

That means our job is changing. We spent decades unsuccessfully sounding the alarm about the coming privacy apocalypse to an indifferent horde, and, just like the environmentalists and the tobacco activists, we failed. The current, dismal state of privacy and technology means that millions have come to harm and more harm is on the way. The only thing worse than all that suffering would be for it all to go to waste.

The people who finally understand that there's a problem don't necessarily understand that there is a solution: we must undertake an overhaul of technology, law, social norms, and business that makes the future fit for human habitation. The processes set in motion by the aggregation and retention of data have unavoidable consequences, and we're not going to stop using email or satellite navigation or social media, so it will be tempting to conclude that the cause is hopeless, and that the only thing to do is surrender to the inevitable.

This is privacy nihilism, and it is the new front for privacy activists. Technology has given us unprecedented, ubiquitous surveillance, but it has also given us unprecedented strong encryption that makes evading surveillance more possible than at any time in history. We have witnessed the rise of digital monopolists who abuse us without fear of being punished by a competitive market, but we've also lived through the rise of digital tools that make it possible for ordinary people to organize themselves and demand strong, vigorously enforced antitrust rules. Our lawmakers foolishly demand bans on working cryptography, but they're also increasingly embarrassed and even politically destroyed by privacy breaches, and they're joining the chorus demanding better privacy rules.

The bad news is that convincing people to pay attention to harms that are a long way off is so hard that it may very well be impossible. But the good news is that convincing people that the disaster they're living through can be averted, through their good will and forceful deeds, is much, much easier. As a species, we may be tragically shortsighted, but we make up for it by being so ferociously indomitable when disaster looms.

It's looming now. Time to start fighting. ●

"The truckers are bringing us the news from the future."

Julia Angwin and Trevor Paglen in Conversation

Moderated by Reyhan Harmanci

Amidst the daily onslaught of demented headlines, two words are often found forming a drumbeat, connecting seemingly disparate news events: security and privacy. Social media is a story of us sacrificing our privacy, knowingly and unknowingly. This sacrifice is often framed as an individual choice rather than collective action. Breaches of banks, voting machines, hospitals, cities, and data companies like Equifax expose the lie of security–namely, that our information is ever safe from intruders. But artist Trevor Paglen and journalist Julia Angwin, two of the world's sharpest minds on those subjects, would rather not use those words.

The Berlin-based Paglen, who holds a doctorate in geography from Berkeley, has built his art career on helping to make the unseen visible–using photography to draw our attention to the cables running under the Atlantic that constitute the internet's infrastructure, for instance, or enlisting amateur trackers to help find American-classified satellites and other unknown space debris.

Angwin, an award-winning journalist and author, recently left ProPublica to build a newsroom that will continue her work of holding tech companies accountable for their impact on society. She has revealed, for instance, the racism and bigotry that undergird Facebook algorithms allowing the company to sell demographics like "Jew hater" to willing advertisers. I talked with Trevor and Julia in New York City about all this and more.

REYHAN HARMANCI: So you guys know each other.

JULIA ANGWIN & TREVOR PAGLEN: Yes.

RH: When did your paths cross?

JA: We met *before* Snowden, yeah, because I was writing my book on surveillance, and you were doing your awesome art.

TP: And you reached out to me and I was like, "You're some crazy person," and didn't write back.

JA: I wrote to him, and I was like, "I want to use one of your art pieces on my website, and how much would it cost?" and he was like, "Whatevs, you're not even cool enough."

TP: That is one of the most embarrassing moments of my life. You know, I get a million of those emails every day.

JA: So do I! I ignore all of them, and then one day that person will be sitting right next to me, and I'll be like, "Heh, sorry."

The problem with email is that everyone in the world can reach you. It's a little too much, right? It's just not a sustainable situation. But yeah, we met sometime in the pre-Snowden era, which now seems like some sort of sepia-toned...

TP: ...wagons and pioneers...

JA: Exactly. And what's interesting about this world is that we'd been walking in the same paths. It's been an interesting journey, just as a movement, right? It started off about individual rights and surveillance and the government, and very nicely moved into social justice, and I think that's the right trajectory. Because the problem with the word formerly known as *privacy* is that it's not an individual harm, it's a collective harm.

TP: That's right.

JA: And framing it around individuals and individual rights has actually led to, I think, a trivialization of the issues. So I'm really happy with the way the movement has grown to encompass a lot more social justice issues.

RH: I want to get into why you guys like the word *trust* better than *surveillance* or *privacy*, two words that I think are associated with both of you in different ways. Do you have any memories of the first time you came to understand that there was an apparatus that could be watching you?

JA: That's an interesting question. I was the kid who thought my parents knew everything I was doing when I wasn't with them, so I was afraid to break the rules. I was so obedient. My parents thought television was evil, so at my friends' houses I would be like, "I can't watch television." And they'd be like, "But this is your one chance!"

And I have spent my whole life recovering from that level of obedience by trying to be as disobedient as possible. I did feel like my parents were an all-seeing entity, you know, and I was terrified of it. And I think that probably did shape some of my constant paranoia, perhaps.

RH: Right, yeah. I was a really good kid, too.

JA: Yeah, it's no good, that goodness.

TP: I think I had the sense of growing up within structures that didn't work for me and feeling like there was a deep injustice around that. Feeling like the world was set up to move you down certain paths and to enforce certain behaviors and norms that didn't work for me, and realizing that the value of this word formerly known as *privacy*, otherwise known as *liberty*, plays not only at the scale of the individual, but also as a kind of public resource that allows for the possibility of, on one hand, experimentation, but then, on the other hand, things like civil liberties and self-representation.

You realize that in order to try to make the world a more equitable place, and a place where there's more justice, you must have sectors of society where the rules can be flexible, can be contested and challenged. That's why I think both of us aren't interested in privacy as a concept so much as anonymity as a public resource. And I think that there's a political aspect to that in terms of creating a space for people to exercise rights or demand rights that previously didn't exist. The classic example is civil rights: there are a lot of people that broke the law. Same with feminism, same with ACT UP, same with the entire history of social justice movements.

And then as we see the intensification of sensing systems, whether that's an NSA or a Google, we're seeing those forms of state power and the power of capital encroaching on moments of our everyday lives that they were previously not able to reach. What that translates into if you're a corporation is trying to extract money out of moments of life that you previously didn't have access to, whether that's my Google search history or the location data on my phone that tells you if I'm going to the gym or not. There is a set of de facto rights and liberties that arise from the fact that not every single action in your everyday life has economic or political consequences. That is one of the things that I think is changing, and one of the things that's at stake here. When you have corporations or political structures that are able to exercise power in places of our lives that they previously didn't have access to, simply because they were inefficient, it adds up to a society that's going to be far more conformist and far more rigid, and actually far more predatory.

JA: I agree.

RH: What makes you guys dislike the word *privacy*?

JA: Well, when I think of privacy, I think of wanting to be alone. I mean, I think that's the original meaning of it. But the truth is that the issues we're discussing are not really about wanting to be solitary. They're actually about wanting to be able to participate in the wonderful connectedness that the internet has brought to the world, but without giving up everything. So privacy, yes, I do want to protect my data and, really, my autonomy, but it doesn't mean that I don't want to interact with people. I think privacy just has a connotation of, "Well you're just, like, antisocial," and it's actually the opposite.

TP: Yeah, absolutely. I think it's right to articulate the collective good that comes through creating spaces in society that are not subject to the tentacles of capital or political power, you know? Similar to freedom of speech, not all of us have some big thing that we want to exercise with our freedom of speech, but we realize that it's a collective good.

JA: Yeah, exactly. And so I think *privacy* feels very individual, and the harm that we're talking about is really a societal harm. It's not that it's not a good proxy; I use that word all the time because it's a shorthand and people know what I'm talking about, especially since I wrote a book with that in the title. But I do feel like it just isn't quite right.

TP: The other side of *privacy* is the word *surveillance*, which certainly is a word that started becoming popular in the '90s when surveillance cameras were installed around cities, and everyone was like, "Oh, you're being watched by some guy in front of a bunch of screens." But those screens are gone, and that guy disappeared and is unemployed now because all of that has been automated. The other aspect of surveillance, I think, is that it has been historically associated with forms of state power, and that's certainly still true. But with that is surveillance capitalism now, which is how you monetize the ability to collect very intimate kinds of information.

JA: Yeah, I think *surveillance* is a slightly better word for what we're talking about, but because of the connotation a lot of people see it only as government surveillance. That doesn't encompass everything, and what it also doesn't encompass is the effect of it. Surveillance is just the *act* of watching, but what has it done to the society, right? What does it do to you? What does it do when there're no pockets where you can have dissident views, or experiment with self-presentation in different ways? What does that look like? That's really just a form of social control, and, like you said, a move towards conformity. And so that is, I think, why *surveillance* itself is not quite an aggressive enough word to describe it.

TP: Because *surveillance* also implies a kind of passivity on the part of the person doing the surveilling, and that's not true anymore; these are active systems.

RH: What does it mean to be an active system?

TP: Super simple example: you're driving in your car and you make an illegal right turn on a red light, and a camera detects that, records your license plate, and issues a traffic ticket–and there's no human in the loop. You can abstract that out to being denied employment based on some kind of algorithmic system that is doing keyword filtering on résumés. A lot of the work you've been doing, Julia, is on this.

JA: Yeah, I mean the thing is, the surveillance is just the beginning of the problem. It's about, once the data is collected, what are they going to do with it? Increasingly, they're making automated decisions about us.

And, weirdly, it seems as though the decisions they're automating first are the most high-stakes human decisions. The criminal justice system is using these systems to predict how likely you are to commit a crime in the future. Or there are systems that are figuring out if you're going to be hired just by scanning résumés. I think it's so funny because the automated systems that most people use are probably maps, and those things suck. Like half the time they're telling you to go to the wrong place, and yet we're like, "I know! This stuff is so awesome; let's use it for really high-stakes decisions about whether to put people in jail or not!"

TP: Or kill them!

RH: I think a lot of times people say, "Well, I'm not doing anything wrong. Take my information, sell me stuff. I am not making bombs in my basement." But that's only seemingly the tip of the iceberg.

JA: Well, that's a very privileged position. Basically, the people who say that to me are white. The number of things that are illegal in this world is really high. We have a lot of laws, and there are certain parts of our community, mostly brown people, who are subject to much stronger enforcement of them. And so I think the feeling of "I'm not doing anything wrong" is actually a canard. Because, of course, even if you were caught doing it, you would get off because that's how our beautiful society is set up. But in fact, people are monitoring the Twitter feeds of Black Lives Matter, and all the Ferguson people, and trying to find anything to get them on. And that is what happens with surveillance–you can get somebody on anything.

When I was working on my book, I went to visit the Stasi archives in Berlin where they have the files that the Stasi kept on the citizens, and I did a public record request. I got a couple of files which are actually translated, and they're on my website. What was so surprising to me was that the Stasi had so little, but they only needed some small thing. Like this one high school student, he skipped class a couple times or something, so they went to his house and were like, "You should become an informant, or else we're going to punish you." It was all about turning people into informants based on a tiny shred of something they had. And I think that's really what surveillance is: it's about the chilling effect, the self-censorship, where you're not willing to try anything risky. There could be very strong consequences, particularly if you're part of a group that they want to oppress.

TP: I disagree in the sense that I think... Okay, Reyhan, you're pregnant. Let's say I have all your geolocation data, and I know that you visited a liquor store on a certain day. Do I sell that information to your health insurance provider? That's going to have consequences for your ability to access health care, or mean you may have to

Source:
National Registry
of Exonerations

3:1

Likeliness that, once stopped, a black driver will be searched compared to a white driver.

△ DATA

pay higher premiums. I think it's not just a kind of legalistic framework; it's a framework about "What are places in which I am able to leverage information in order to extract money from you, as well as determine whether or not you become part of the criminal justice system?"

RH: Sometimes I have trouble imagining that all of this is happening.

JA: It does seem unreal. In fact, what I find so shocking is that I felt like I had tried to paint the most pessimistic view I could in 2014, and it's so much worse; it's so much worse than I imagined. So, let's look at one simple thing in this country, and then one in China.

In this country, these risk scores that I've been talking about are being used when you get arrested. They give you a score, 1 to 10, of how likely you are to commit a future crime, and that can determine your conditions of bail. There's this huge movement across the nation to, instead of putting people in jail–the higher-risk people–just put them on electronic ankle bracelets. So essentially an automated decision forces somebody to literally have a GPS monitor on them at all times. That is something I actually didn't think would happen this fast. Three years ago I would have said, "Nah, that's a little dystopian."

Now think about China: there's a region that is very Muslim, and the Chinese government is trying to crack down because they think they're all terrorists. I was just with this woman from Human Rights Watch who works in that region and aggregated all the information about which surveillance equipment they'd bought in that region and what they're doing. And they built an automated program, which basically says if you go to the gas station too many times, or you buy fertilizer, or you do anything suspicious. Something pops up that says you're high-risk and they send you to reeducation camp. It's an automated system to determine who gets sent into reeducation camp. And I was like, "Oh my god, I can't believe this is happening *right now*; we're not even talking about the future!"

TP: This Chinese citizen credit score system has been in the news a little bit.

A Sampling of Questions from the COMPAS Risk-Assessment Survey

CRIMINAL HISTORY	How many prior juvenile felony offense arrests?	○ 0 ○ 2 ○ 4 ○ 1 ○ 3 ○ 5+
	How many prior felony (misdemeanor, family violence, sex, weapons) offense arrests as an adult?	○ 0 ○ 2 ○ 4 ○ 1 ○ 3 ○ 5+
NON-COMPLIANCE	How many times has this person violated his or her parole?	○ 0 ○ 2 ○ 4 ○ 1 ○ 3 ○ 5+
	How many times has this person failed to appear for a scheduled criminal court hearing?	○ 0 ○ 2 ○ 4 ○ 1 ○ 3 ○ 5+
FAMILY CRIMINALITY	Was your father (or father figure who principally raised you) ever arrested, that you know of?	○ No ○ Yes
	Did a parent or parent figure who raised you ever have a drug or alcohol problem?	○ No ○ Yes
PEERS	How many of your friends/acquaintances have ever been arrested?	○ 0 ○ 2 ○ 4 ○ 1 ○ 3 ○ 5+
	Have you ever been a gang member?	○ No ○ Yes
RESIDENCE/ STABILITY	Do you have a regular living situation?	○ No ○ Yes
	How often have you moved in the last twelve months?	○ 0 ○ 1 ○ 2+
	Do you live with friends?	○ No ○ Yes
SOCIAL ENVIRONMENT	Is there much crime in your neighborhood?	○ No ○ Yes
	In your neighborhood, have some of your friends or family been crime victims?	○ No ○ Yes

● COMPAS, created by the private company Northpointe Inc., 0059
is one of the most widely used risk-assessment softwares.
The responses to its 137-question survey are used to
calculate a defendant's risk score.

EDUCATION	Were you ever suspended or expelled from school?	○ No ○ Yes
	Did you fail or repeat a grade level?	○ No ○ Yes
	Did you complete your high school diploma or GED?	○ No ○ Yes
VOCATION (WORK)	Have you ever been fired from a job?	○ No ○ Yes
	Right now, do you feel you need more training in a new job or career skill?	○ No ○ Yes
	Do you often have barely enough money to get by?	○ No ○ Yes
LEISURE/RECREATION	Do you often feel bored with your usual activities?	○ No ○ Yes
	Do you feel discouraged at times?	○ No ○ Yes
	How much do you agree or disagree with these statements:	I have a best friend I can talk about with everything. ○ Strongly disagree ○ Disagree ○ Agree ○ Strongly agree I feel lonely. ○ Strongly disagree ○ Disagree ○ Agree ○ Strongly agree
CRIMINAL PERSONALITY	*How much do you agree or disagree with these statements:*	I am seen by others as cold and unfeeling. ○ Strongly disagree ○ Disagree ○ Agree ○ Strongly agree I always practice what I preach. ○ Strongly disagree ○ Disagree ○ Agree ○ Strongly agree I'm really good at talking my way out of problems. ○ Strongly disagree ○ Disagree ○ Agree ○ Strongly agree

▽ **Prediction Failures of COMPAS' Algorithm**

Respondents who were labeled higher-risk but didn't re-offend:

| Black: 44.9% | White: 23.5% |

Respondents who were labeled lower-risk but did re-offend:

| Black: 28.0% | White: 47.7% |

Basically the pilot programs, which are in operation in advance of the 2020 national rollout, are able to monitor everything that you do on social media, whether you jaywalk or pay bills late or talk critically about the government, collect all these data points, and assign you a citizen score. If you have a high one, you get, like, discounts on movie tickets, and it's easier to get a travel visa, and if you have a lower one, you're going to have a really bad time. It's creating extreme enforcement mechanisms, and I think a lot of us here in the U.S. say, "Oh well, that's this thing that can only happen in China because they have a different relationship between the state and the economy and a different conception of state power." But the same exact things are happening here, they're just taking a different shape because we live under a different flavor of capitalism than China does.

RH: Do Chinese citizens know about this?

JA: Yes.

RH: So these scores are not secret or subterranean…

JA: No, they're meant to be an incentive system, right? It really is just a tool for social control; it's a tool for conformity, and probably pretty effective, right? I mean, you get actual rewards; it's very Pavlovian. You get the good things or you don't. It's just that governments have not had such granular tools before. There have been ways that the government can try to control citizens, but it's *hard* trying to control a population! These tools of surveillance are really part of an apparatus of social control.

RH: How good is the data?

JA: That's the thing: it doesn't matter. First of all, the data all sucks, but it doesn't matter. I was asking this analyst from Human Rights Watch why they don't just put any random person in the detention camp, because it's really just meant as a deterrent for everyone else to be scared. You don't really need any science. But what they realize is that they're actually trying to attain the good behavior, so even though it cost money to build this algorithm, which nobody's testing the accuracy of, it creates better incentives. People will act a certain way. So the data itself could be garbage, but people will think, "Oh, if I try to do good things, I will win. I can game the system." It's all about appealing to your illusion that you have some control. Which is actually the same thing as the Facebook privacy settings—it's built to give you an illusion of control. You fiddle with your little dials, and you're like, "I'm so locked down!" But you're not! They're going to suggest your therapist as your friend, and they're going to out you to your parents as a gay person. You can't control it!

RH: I think I have bad personal habits, because I have trouble believing that my phone is really giving information to Google or Facebook or whatever about where I am. I have trouble believing that on some level.

TP: It tells you!

RH: But that I matter enough, even as a consumer, to them.

JA: Well you don't really matter to them; I mean, they need everybody. It's more about the voraciousness. They need to be the one place that advertisers go, need to be able to say, "We know everything about *everyone*." So it's true, they're not sitting around poring over the Reyhan file—unless you have somebody who hates you there who might want to pore over that file. But it's more about market control. It's similar to state control in the sense that, in order to win, you need to be a monopolist. That's actually kind of just the rule of business. And so they—Google and Facebook—are in a race to the death to know everything about everyone. That's what they compete on, and that's what their metric is. Because they go to the same advertisers and they say, "No, buy with us." And the advertisers are like, "Well, can you get me the pregnant lady on the way to the liquor store? Because that's what I want."

And whoever can deliver that, they're in. They're in a race to acquire data, just the way that oil companies race to get the different oil fields.

TP: The one thing I want to underline again is that we usually think about these as essentially advertising platforms. That's not the long-term trajectory of them. I think about them as, like, extracting-money-out-of-your-life platforms.

JA: The current way that they extract the money is advertising, but they are going to turn into–are already turning into–intermediaries, gatekeepers. Which is the greatest business of all. "You can't get anything unless you come through us."

TP: Yeah, and health insurance, credit…

JA: That's why they want to disrupt all those things–that's what that means.

RH: So how is that going to play out? Is it like when you sign up for a service and you can go type in your email or hit the Facebook button? Is that how it's happening?

JA: Well, that's how it's happening right now. But you know Amazon announced health care, right? We're going to see more of that. My husband and I were talking recently about self-driving cars. Google will also then become your insurer. So they're going to need even more data about you, because they're going to disrupt insurance and just provide it themselves. They know more about you than anyone, so why shouldn't they pool the risk? There's a lot of ways that their avenues are going to come in, and to be honest, they have a real competitive advantage because they have that pool of data already.

And then also it turns out, weirdly, that no one else seems to be able to build good software, so there is also this technological barrier to entry, which I haven't quite understood. I think of it like Ford: first, when he invented the factory, he had a couple decades' run of that being *the* thing. And then everyone built factories, and now you can just build a factory out of a box. So I think that that barrier to entry will disappear. But right now, the insurer companies–even if they had Google's data set–wouldn't actually be able to build the software sophisticated enough. They don't have the AI technology to predict your risk. So these tech companies have two advantages, one of which might disappear.

TP: But the data collection is a massive advantage.

JA: It's a *massive* advantage.

TP: I mean, a lot of these systems simply don't work unless you have data.

JA: Yes, and sometimes they talk about it as their "training data," which is all the data that they collect and then use to train their models. And their models, by definition, can make better predictions because they have more data to start with. So, you know, when you look at these companies as monopolies, some people think of it in terms of advertising revenue. I tend to think about it in terms of the training data. Maybe *that* is what needs to be broken up, or maybe it has to be a public resource, or something, because that's the thing that makes them impossible to compete with.

RH: How good are they at using this data? A lot of times I've been very unimpressed by the end results.

JA: Well, that's what happens with monopolies: they get lazy. It's a sure sign of a monopoly, actually.

TP: Yeah, how good is your credit reporting agency?

JA: It's garbage!

RH: Totally! It all seems like garbage, you know? Facebook is constantly tweaking what they show me, and it's always wrong.

JA: Right. That's why there was an announcement recently, like, "Oh, they're going to change the news feed." They don't care what's in the news feed. It doesn't matter, you're not the customer. They don't want you to leave, but you also have nowhere to go, so...

RH: Right. And it's like Hotel California: you'll never actually be able to leave.

JA: Yes, correct!

RH: So they're engaged in this massive long-term effort to be the gatekeepers between you and any service you need, any time you need to type in your name anywhere.

JA: I mean, really, the gatekeeper between you and your money. Like, whatever it takes to do that transaction.

RH: Or to get other people's money to them.

JA: Yeah, sure, through you.

TP: Let's also not totally ignore the state, too, because records can be subpoenaed, often without a warrant. I think about Ferguson a lot, in terms of it being a predatory municipal system that is issuing, like, tens of thousands of arrest warrants for a population of, like, that many people, and the whole infrastructure is designed to just prey on the whole population. When you control those kinds of large infrastructures, that's basically the model.

RH: And that is very powerful. How does the current administration's thirst for privatization play into this sort of thing? Because I could imagine a situation

Arrest Warrants in Ferguson, Missouri

Population of Ferguson, Mo., in 2014: 21,069

16,000+

As of December 2014, over sixteen thousand people had outstanding arrest warrants.

4%

96%

In Ferguson, Mo., from 2012–2014, African Americans accounted for 96 percent of known arrests made exclusively because of an outstanding municipal warrant.

Source: United States Department of Justice Civil Rights Division

where you want to have national ID cards, and who better to service that than Facebook?

JA: Oh, that effort is underway. There is a national working group that's been going on for five years, actually, that's looking for the private sector to develop ways to authenticate identities, and the tech companies are very involved in it. The thing about this administration that's interesting is, because of their failure to staff any positions, there is some...

RH: A silver lining!

JA: Yeah, some of these efforts are actually stalled! So that actually might be a win, I don't know. But the companies have always been interested in providing that digital identity.

TP: We don't just have to look at the federal level. You can look at the municipal level, in terms of cities partnering with corporations to issue traffic tickets. Or look at Google in schools. I do think, with the construction of these infrastructures, there is a kind of de facto privatization that's happening. Apple or Google will go to municipalities; they will provide computers to your schools, but they're going to collect all the data from all the students using it, et cetera. Then they'll use that data to in other ways partner with cities to charge for services that would previously have been public services. Again, there are lots of examples of this with policing–take Vigilant Solutions. This company that does license plate reading and location data will say, "Okay, we have this database of license plates, and we're going to attach this to your municipal database so you can use our software to identify where all those people are who owe you money, people who have arrest warrants and outstanding court fees. And we're going to collect a service charge on that." That's the business model. I think that's a vision of the future municipality in every sector.

JA: You know, that's such a great example because Vigilant really is... basically, it's repo men. So essentially it starts as predatory. All these repo men are already driving around the cities all night, looking for people's cars that the dealers are repossessing because people haven't paid. Then they realized, "Oh, the government would actually really love a national database of everybody's license plates," so they started scanning with these automated license plate readers on their cars. So the repo men across the country go out every night—I think it was Baltimore where Vigilant's competitor had forty cars going out every night—taking photographs of where every car is parked and putting them in a database. And then they build these across the nation, and they're like, "Oh, DHS, do you want to know where a car is?" and then they bid for a contract. And so you add the government layer and then you add back the predatory piece of it, which is... really spectacular.

RH: But if you talk to an executive at Vigilant, they would be like, "What's the problem? We're just providing a kind of service."

JA: Yeah, I did spend a bunch of time with their competitors based in Chicago, met a bunch of very nice repo men who said, "Look, we're doing the Lord's work. I mean, these are bad people, and we have to catch them." And then I said, "Can you look up my car?" and *boom*, they pushed a button, and there it was, right on the street in front of my house! I was like, holy shit.

TP: Wow.

JA: That is so creepy.

RH: That's crazy! That's the kind of thing where, again, I'd be like, "Nobody knows where my car is; *I* don't know where my car is! They're not fucking finding it," you know?

JA: Right.

TP: But you could take it further. Think about the next generation of Ford cars. They know exactly where your car is all the time, so now Ford becomes a big data company providing geolocation data.

RH: And you're not even going to own your car since its software is now property of the carmaker.

TP: It goes back to the thing about whether you're going to liquor stores, whether you're speeding. Tomorrow, Ford could make a program for law enforcement that says, "Every time somebody speeds, whatever jurisdiction you're in, you can issue them a ticket if you sign up for our..." You know? I mean, this is a full traffic enforcement partnership program.

JA: And one thing I just want to say, because I have to raise the issue of due process at least once during any conversation like this. My husband and I fight about this all the time: he's like, "I want them to know I'm driving safely because I'm safe, and I'm happy to have the GPS in the car, and I'm awesome, and I'm going to benefit from this system." But the problem is, when you think about red-light cameras and automated tickets, which are already becoming pretty pervasive, it's really a due process issue. How do you fight it? It would probably be really hard to do. Like, are we going to have to install dashboard cameras surveilling ourselves in order to have some sort of protection against the state surveilling us, or whoever is surveilling us? I think a lot of the issues raised by this type of thing are really about, and I hate to say it, but our individual rights. Like about the fact that you can't fight these systems.

TP: That's true. There's a lot of de facto rights we've had because systems of power are inefficient. Using a framework of criminal justice where this is the law, this is the speeding limit, and if you break that you're breaking the law–that's not actually how we've historically lived.

JA: Right.

TP: And that is very easy to change with these kinds of systems.

RH: Yeah, because then if you're looking to target somebody, that's where the data's weaponized. And also, there's so much rank incompetence in criminal justice, and in all these human systems. Maybe its saving grace has been that humans can only process so much stuff at any one time.

TP: I think the funny thing is that it's been police unions most vocally opposed to installing things like red-light cameras because they *want* that inefficiency.

RH: Yeah, and they say, "We're making judgment calls."

JA: I mean there was this crazy story–you know the truckers, right? The truckers are being surveilled heavily by their employers. A lot of them are freelance, but there are federal rules about how many hours you can drive. It's one of these classic examples–sort of like welfare, where you can't actually eat enough with the amount of money they give you, but it's illegal to work extra. So similarly, for truckers, the number of hours you need to drive to make money is more than the hours you are allowed to drive. And to ensure the truckers don't exceed these miles, they've added all these GPS monitors to the trucks, and the truckers are involved in these enormous countersurveillance strategies of, like, rolling backwards for miles, and doing all this crazy stuff. And there was this amazing thing one day at Newark, a year or two ago, where the control tower lost contact with the planes at Newark Airport, and it turned out it was because of a trucker. He was driving by the airport and he had such strong countersurveillance for his GPS monitor that he interrupted the air traffic!

TP: Wow!

RH: Wild. God, a decade ago I was on BART, and I never do this, but I was talking to the guy next to me and he was a truck driver. He was telling me

AMAZON

Whole Foods
Grocery

Zappos
Clothing

PillPack
Prescription
Fulfillment

Ring
Smart-home
tech

ALPHABET

Google
All Google products, including
smart home, social media, GPS,
communication, virtual reality, audio,
photo, video, travel, news

Verily
Health care and
disease research

Calico
Life-extension
research

FACEBOOK

Ascenta
Drones

ProtoGeo Oy
Health and fitness
monitoring/location
tracking

Oculus VR
Virtual reality

● A non-exhaustive look at the cross-industry subsidiary companies owned by three of the biggest tech giants.

0071

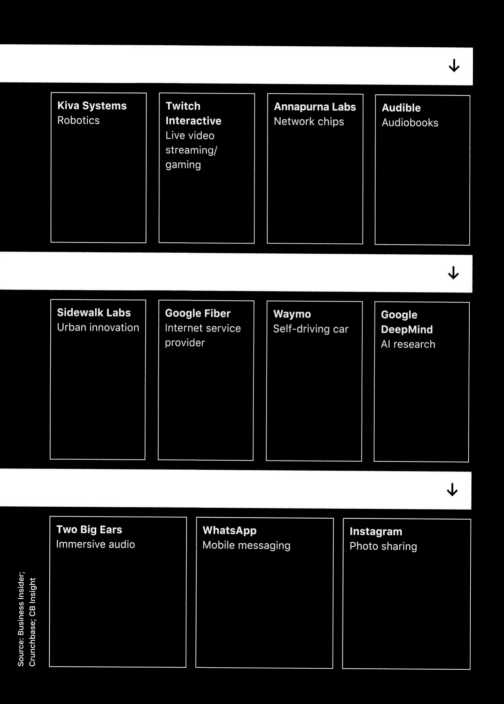

↓

Kiva Systems
Robotics

Twitch Interactive
Live video streaming/ gaming

Annapurna Labs
Network chips

Audible
Audiobooks

↓

Sidewalk Labs
Urban innovation

Google Fiber
Internet service provider

Waymo
Self-driving car

Google DeepMind
AI research

↓

Two Big Ears
Immersive audio

WhatsApp
Mobile messaging

Instagram
Photo sharing

Source: Business Insider; Crunchbase; CB Insight

about the surveillance of it at the time. He was like, "This has already been going on." This was ten years ago! He said the first kinds of self-driving cars are going to be trucks.

JA: Yes, that's right, they are going to be trucks.

RH: I just remember getting off BART and being like, "What if that was an oracle or something?"

JA: The truckers are bringing us the news from the future.

RH: So this conversation is being framed as "the end of trust," but did we *have* trust?

JA: I'm not sure I agree with "the end of trust." I'm more into viewing this as an outsourcing of trust right now. So essentially I think that in this endless quest for efficiency and control we've chosen–and sometimes I think it's a good thing–to trust the data over human interactions. We're in this huge debate, for instance, about fake news. We've decided it's all Facebook's problem. They need to determine what is real and what is not. Why did we outsource the trust of all of what's true in the world, all facts, to Facebook? Or to Google, to the news algorithm? It doesn't make any sense.

And so in the aggregation of data, we have also somehow given away a lot of our metrics of trust, and that's partly because these companies have so much data, and they make such a strong case that they can do it with the math, and they're going to win it with the AI, they're going to bring us the perfect predictions. What I'm really concerned about is the fact that we've broken the human systems of trust that we understood. Studies show that when you look at a person, you establish trust with them, like, in a millisecond. And you can tell people who are not trustworthy right away. You can't even articulate it; you just know. But online and through data, we just don't have those systems, so it's really easy to create fake stuff online, and we don't know how to build a trustworthy system. That's what is most interesting to me about the moment we're in.

I think we're at a moment where we could reclaim it. We could say, "This is what journalism is, and the rest of that stuff is not." Let's affirmatively declare what's trustworthy. We've always had information gatekeepers, right? Teachers would say, "Read this encyclopedia, that's where the facts are." Now my kids come home and just look things up on the internet. Well, that's not the right way to do it. I think we're at a period where we haven't figured out how to establish trust online, and we've outsourced it to these big companies who don't have our interests at heart.

RH: In terms of "the end of trust," what *does* feel threatening or worth creating collective action around?

TP: For me, it's the ability of infrastructures, whether they're private or policing, to get access to moments of your everyday life far beyond what they were capable of twenty years ago. It's not a change in kind, it's just the ability for capital and the police to literally colonize life in much more expansive ways.

RH: Both of you guys are engaged in work that tries to make this stuff visible–and it's very hard. The companies don't want you to see it. A lot of your work, Trevor, has been about the visceral stuff, and your work, Julia, has been about saying, "We've tested these algorithms, we can show you how Facebook chooses to show a twenty-five-year-old one job ad and a fifty-five-year-old another." What do you wish you could show people?

TP: There're very clear limits on the politics of making these things visible. Yeah, sure, it's step one, and it's helpful because it develops a vocabulary that you can use to have a conversation. The work that you do, Julia, creates a vocabulary for us to be able to talk about the relationship between Facebook and due process, which we didn't have the vocabulary, really, to even have an intelligible conversation about before.

I think it's just one part of a much bigger project, but doesn't in and of itself do anything. That's why we have to develop a different sense of civics, really. A more conscious sense of which parts of our everyday life we want to subject to these kinds of infrastructures, and which we don't. There is good stuff that you can do, for example, with artificial intelligence, like encouraging energy efficiency. So I think we need to say, "Okay, these are things that we want to optimize, and these are things that we don't want to optimize." It shouldn't just be up to capital, and up to the police, and up to the national security state to decide what those optimization criteria are going to be.

JA: I strongly believe that in order to solve a problem, you have to diagnose it, and that we're still in the diagnosis phase of this. If you think about the turn of the century and industrialization, we had, I don't know, thirty years of child labor, unlimited work hours, terrible working conditions, and it took a lot of journalist muckraking and advocacy to diagnose the problem and have some understanding of what it was, and then the activism to get laws changed.

I feel like we're in a second industrialization of data information. I think some call it the second machine age. We're in the phase of just waking up from the heady excitement and euphoria of having access to technology at our fingertips at all times. That's really been our last twenty years, that euphoria, and now we're like, "Whoa, looks like there're some downsides." I see my role as trying to make as clear as possible what the downsides are, and diagnosing them really accurately so that they can be solvable. That's hard work, and lots more people need to be doing it. It's increasingly becoming a field, but I don't think we're all the way there. One thing we really haven't tackled: it sounds really trite, but we haven't talked about brainwashing. The truth is that humans are really susceptible. We've never before been in a world where people could be exposed to every horrible idea in the universe. It looks like, from what you see out there, people are pretty easily convinced by pretty idiotic stuff, and we don't know enough about that, that brain psychology. What's interesting is the tech companies have been hiring up some of the best people in that field, and so it's also a little disturbing that they may know more than the academic literature that's presented.

RH: That's crazy.

JA: I know, I hate saying *brainwashing*. There's probably a better word for it. *Cognitive persuasion* or something.

RH: Or just understanding humans as animals, and that we're getting these animal signals.

JA: Yeah, right. I mean, we have those behavioral economics. That's led a little bit down the road, but they just haven't gone as far as what you see with the spiral of the recommendation engines, for instance, on Facebook, where you search for one thing related to vaccines and you're led down every single conspiracy theory of anti-vaxx and chemtrails, and you end up drinking raw water out of the sewer at the end of it.

RH: Right, you spend forty dollars at a Rainbow Grocery in San Francisco buying your raw water.

JA: It's an amazing pathway. People do get radicalized through these types of things, and we just don't understand enough about countermeasures.

RH: Yeah, that's really scary. Do you guys interact with young people? Do they give you any sense of hope?

JA: My daughter's awesome. She's thirteen. I actually have great hope for the youth because she really only talks on Snapchat, so she's completely into ephemeral everything. And her curation of her public personality is so aggressive and so amazing. She has, like, five pictures on her Instagram, heavily curated, and she has, of course, five different Instagrams for different rings of friendship. It's pretty impressive! I mean it takes a lot of time, and I'm sad for her that she has to spend that time, because at that age literally all I wore was pink and turquoise for two straight years and there's no documentation of it. Like, nothing. It was the height of Esprit. And I can just tell you that, and you'll never see it. She will always have that trail, which is a bummer, but I also appreciate that she's aware of it and she knows how to manage it even though it's a sad task, I think.

RH: Yeah. I'm always fascinated by teenagers' computer hygiene habits.

JA: I think they're better than adults'.

RH: I'm curious about AI stuff. There's a lot of hype around AI, but all I ever really see of it are misshapen pictures on Reddit or something. I'm like, "We're training them to recognize pictures of dogs," you know? What's the end goal there? Is this technology going to powerfully shape our future in a way that is scary and hidden? Or is it a lot of work, a lot of money, and a lot of time for a result that seems, frankly, not that impressive?

TP: Yeah, there're misshapen pictures on Facebook and Reddit and that sort of thing–that's people like us playing with it. That's not actually what's going on under the hood. That's not what's driving these companies.

JA: I want us to back up a little bit because AI has been defined too narrowly. AI used to just mean automated decision-making. Then recently it's been defined mostly as machine learning. Machine learning is in its early stages, so you see that they make mistakes. Google characterizing black faces as gorillas because they had shitty [racially biased] training data. But in fact the threat of AI is, to me, broader than just the machine learning piece of it. The threat is really this idea of giving over trust and decision-making to programs that we've put some rules into and just let loose. And when we put those rules in, we put our biases in. So essentially, we've automated our biases and made them inescapable. You have no due process anymore to fight against those biases.

One thing that's really important is we're moving into a world where some people are judged by machines, and that's the poor and the most vulnerable, and some people are judged by humans, and that's the rich. And so essentially only some people are sorted by AI–or by whatever we call it: algorithms, automated decisions–and then other people have the fast track. It's like the TSA PreCheck.

TP: An article came out reporting that Tesco, the UK's largest private employer, uses AI as much as possible in its hiring process. Whereas this afternoon I had to be part of a committee hiring the new director of our organization: it's all people.

JA: Yes, right, exactly. It's almost like two separate worlds. And that's what I think is most disturbing. I have written about the biases within AI's technology, but the thing is, those biases are only within the pool of people who are being judged by it, which is already a subset.

TP: My friend had a funny quote. We were talking about this and she said, "The singularity already happened: it's called neoliberalism."

JA: Oh yeah, that's the greatest quote of all time. I will say this, just because I want to be on the record about it: I really have no patience for singularity discussions because what they do–just to be really explicit about it–is move the ball over here, so that you're not allowed to be concerned about anything that's happening right now in real life. They're like, "You have to wait until the robots murder humans to be worried, and you're not allowed to be worried about what's happening right now with welfare distribution, and the sorting of the 'poor' from the 'deserving poor,'"–which is really what all of those automated systems are: this idea that there are some people who are deserving poor and some people who aren't. Morally, that is already a false construct, but then we've automated those decisions so they're incontestable.

RH: So you could go through the whole system and never see a human, but all the human biases have already been explicitly embedded.

JA: Yeah, there's a great book just out on that called *Automating Inequality* by Virginia Eubanks that I highly recommend.

RH: Can you two talk a bit about your personal security hygiene?

TP: Well, people are always like, "We choose to give our data to Google." No, we don't. If you want to have your job, if you want to have my job, you have to use these devices. I mean, it's just part of living in society, right?

JA: Correct.

TP: I have somebody who works for me who chooses not to, and it's a complete, utter headache where literally the whole studio has to conform to this one person who won't have a smartphone. It's worth it because they're really talented, but at the same time, at any other company, forget it. These are such a part of our everyday infrastructures that you don't have the individual capacity to choose to use them or not. Even if you don't use Facebook, they still have your stuff.

JA: Yeah, I agree. I don't want to give the impression that there is some choice. The reason I wrote my book from the first-person account of trying to opt out of all these things was to show that it didn't work. And I'm not sure everybody got that picture, so I want to just state it really clearly. I do small measures here and there around the corners, but, no, I live in the world, and that's what the price of living in the world is right now. I want to end on the note that this is a collective problem that you can't solve individually.

RH: Right, like, downloading Signal is great, but it's not going to solve the larger issues.

TP: Exactly. I think a really important takeaway from this conversation is that these issues have historically been articulated at the scale of the individual, and that makes no sense with the current infrastructures we're living in.

JA: Right. But do download Signal. ●

Should Law Enforcement Use Surveillance?

Myke Cole, Hamid Khan, & Ken Montenegro

YES, LAW ENFORCEMENT SHOULD practice surveillance, but it needs to do so accountable to a furious public that rightly doesn't trust the police.

Surveillance, both electronic and physical, is a critical tool for police. Skilled criminals succeed precisely because they have made a study of police capabilities, and work to reduce the possibility of detection. Undermanned police forces are unable to be everywhere a crime might occur. Targeting surveillance where there is reasonable suspicion of criminal activity can help stop crime.

But that is surveillance used correctly. And right now, it isn't used that way. From Stingray use in New York City, to FBI overreach in seeking a back door into Apple devices, to xenophobic surveillance of Muslims following 9/11, cops have worked overtime to degrade public trust. Set against the backdrop of wanton police killings of unarmed black men across the country, police have squandered the public faith, calling all police practices, even those that are effective and necessary, into question.

Sir Robert Peel's "policing by consent" envisioned ethical police who were granted powers by a public whose trust they worked assiduously to maintain. The NYPD fields somewhere in the range of thirty thousand uniformed officers, supported by approximately twenty thousand civilians, to police a city approaching nine million. If nine million people do not consent to policing by fifty thousand, there is absolutely nothing the police can do.

Peelian consent relies on police-public rapport, which is built in two ways: (1) Police are held accountable, and prosecuted, when they violate the law. (2) Police prove they are effective in making communities safer. To accomplish the first, we need to publicly hold police to account when they kill unarmed black people, overstep authority, and engage in corrupt practices. To accomplish the second, we need every tool in the law enforcement arsenal, including correct and accountable surveillance, untarnished by bias.

The solution is not to abolish surveillance, but to surveil correctly and accountably. To understand that the maintenance of Peelian consent is critical, and worth sacrificing some, but not all, advantages in the fight against crime. A new model is needed, where an engaged public is informed of the types, extent, and also expected limits of police surveillance, and understands why it is necessary. And when surveillance is improperly used, an angry public must not be met with a blue wall of silence, but by a reckoning that sees those responsible for its misuse swiftly and publicly punished.

OPENING
▷ Con: Hamid Khan
& Ken Montenegro

ASKING IF LAW ENFORCEMENT should practice surveillance belies the right question: it's like asking if soldiers should use chemical weapons during wartime instead of asking why we resolve differences with war. If we avoid questions about the history and role of policing, we end up in discussions about how much poison is lethal instead of interrogating why we are taking poison to start with.

Police surveillance is integral to building systems of knowledge and structures of power that preserve and sustain white supremacy and white privilege. As Dr. Simone Browne outlines, the history of surveillance is the history of anti-blackness. Surveillance is yet another practice built upon the lineage of slave patrols, lantern laws, Jim Crow, Red Squads, the war on drugs, the war on crime, the war on gangs, and now the war on terror. Surveillance in policing has never been applied in a neutral manner; it has always been weaponized against those at the margins and those fighting for liberation.

The right question is whether police surveillance is a life-affirming or a life-denying tool. It is elemental to inquire into the history of policing, how policing practices have evolved, and who has experienced the greatest contact with the police. It is also important to demand that resources currently spent creating a harm-producing surveillance state be shifted to life-affirming entities like youth programs, mental health programs, housing, and education, among others.

Once we have named the intent to cause harm baked into police surveillance, our next line of inquiry should be whether it even works. According to the National

Academy of Sciences, attempts to preempt crime through behavioral surveillance and data mining are fundamentally flawed. Furthermore, the threshold for data entered into police databases is extremely low. Police surveillance is flawed by design: one of many examples is the inclusion of infants in the CalGang database.

Contemporary surveillance is highly weaponized surveillance. A portion of surveillance tools used by police comes from counterinsurgency and the other portion comes from the commercial sector. One sector takes life and the other sector reduces valuable experiences and relationships to monetized data points. These weapons often journey from distant war zones like Iraq and Afghanistan to local points of conflict such as the United States' southern border, Ferguson, Baltimore, and Standing Rock.

Law enforcement should not practice surveillance because surveillance is flawed by design due to its sordid historic roots in harm creation, abuse, anti-blackness, and racism.

REBUTTAL
▷ Pro: Myke Cole

EVERYTHING YOU'VE SAID IS correct. Surveillance is put to use in perpetuating systems of white supremacy. It is weaponized against those living at the margins of society. The threshold for initiating surveillance is too low, its practice sloppy and overreaching. The data gleaned from surveillance isn't properly secured.

And it also works. Used correctly, it can protect the lives of those at the margins. It can deter actions of bigotry. It can tear down white supremacy and help marginalized people live in safety.

Though misused by the NRA, Seneca's words still hold true: "A sword kills no one. It is a tool in the hands of a killer." Surveillance is just that–a neutral tool. Tools capable of oppression and ending life were deployed to free my family from Nazi concentration camps. When I deployed for the Deepwater Horizon oil spill, I watched gunboats capable of leveling towns used to soak up toxic chemicals, the warriors manning them engaged in scrubbing oil-soaked marine life clean. Used correctly, tools can fly in the face of all the wrong they've caused. They can even reverse it.

Marginalized communities need effective police protection. Those living below the poverty line are victims of violent crime at more than double the rate of those with higher incomes.

Surveillance can work. An intercepted email prevented the 2009 bombing of the New York City subway. The NSA has done much to earn the derision and mistrust of the American public, but I also worked with them. When General Keith Alexander makes the claim that SIGINT intercepts have disrupted fifty-four separate attack plots, I believe him.

We had a banner in our ready room at Station New York. "Failure is unforgettable," it read. "Success, invisible." Police worry about "black swans"– high-impact, low-probability events. We deploy gunboats to escort the Staten Island Ferry in the highly secure, peaceful Port of New York. When friends would laugh at this, I would always answer: "9,999 times, you're right. The 10,000th time, some idiot packs a Zodiac full of TATP and runs it into the side of the ferry. Now I have over a thousand people in the water. If it's winter, I could scramble every maritime asset in the port, and at least one hundred of them are going to die."

Police have to get it right. Each time. Every time. The communities suffering from police abuses also would have ridden the subway in 2009, and the ferry when that 10,000th time came up.

Reforming police cannot occur at the price of rendering them ineffective.

REBUTTAL
▷ Con: Hamid Khan
& Ken Montenegro

IT'S GOOD THAT THE other side, which has been speaking for law enforcement, concedes that surveillance is a tool which perpetuates white supremacy because then we can ask key questions, such as: Is the damage surveillance causes acceptable? Does it really work? Who dictates acceptable failure rates, and what does "effective" policing look like? The answers point to the truth that the entirety of surveillance–including broken-windows policing, the war on "gangs," the war on "terror"–is designed to cause harm.

Let's start by asking what damage is acceptable, to whom, and why: native, black, brown, and poor communities experience the harm of police surveillance in genocidal proportions. When we talk about surveillance, their experiences and voices should be amplified and considered more than the hyperbole spoken by "undermanned" police keeping us safe from hordes of the "other." Then, the question of whether it works: the NYPD does not impede the crimes on Wall Street because it's too busy engaging in stop-and-frisk. Statistics show that police do not solve crimes in black and brown communities but do endanger those communities in emotional and physical ways–including the constant threat of violence and the actual violence communities of color experience. The city of Miami solves less than a third of all murder cases in many of its black neighborhoods, and when the LAPD was lackadaisically investigating the serial killer the Grim Sleeper it told one activist that he was "only killing hookers." What seems to be acceptable is actually criminalization through racially targeted practices, such as driving while black/brown, and harm creation disguised as protection, like stop-and-frisk or "random" airport searches. With white folks being the only population that's largely safe from the harm of policing, how is this acceptable? Let's recall that the LAPD used the designation NHI to denote black and brown domestic-violence calls. NHI stands for "no humans involved."

Ontologically, surveillance cannot be neutral because it is born of and enables white supremacy. To assert that surveillance is neutral is as inaccurate, counterintuitive, and dangerous as saying a firearm is an instrument of peace. Today, we can look at "predictive" policing and see that it is racist pseudoscience disguised as science. Therefore, to insist that surveillance "works," after conceding that it advances white supremacy, acknowledges that surveillance and policing work only to impede the life possibilities of non-whites and other folks at the margins.

The calculus law enforcement uses to justify surveillance desperately reaches for "because we can't fail" or "only fail once" to excuse the harm.

The interconnected web of surveillance, which we call the Stalker State, fails repeatedly yet is never accountable when it disrupts and steals black and brown lives. While it's easy to tug at heartstrings and claim weapons (including surveillance) liberated the world from the horror of the Nazis, it is important to think of the refrain "cops and Klan work hand in hand" to remember that one does not need to wear a swastika to advance white supremacy. For instance, a seemingly benevolent process like the census was used against the Japanese-American community during internment.

Finally, any discussion of surveillance begs us to examine the violent role of the state and its agents. Surveillance is just one of many oppressive, violent, and deadly tools of law enforcement. If the goal is to create harm, then yes, surveillance is a valuable tool. And if the efficacy of the police is correlated to an intent to cause harm, then yes, surveillance is essential.

We prefer a world where the lives of people of color and other traditionally targeted communities are valued more than police "efficiency."

CLOSING
▷ Pro: Myke Cole

WHEN SYSTEMS FAIL, THE temptation to abandon them is strong. Sometimes, the tooth is too rotten to fill. It must be pulled out.

With surveillance, that is not the case. It has been horribly misused. It has been unaccountable. It has been employed toward harmful ends.

We can make it do better. We can make it serve those it once unjustly targeted.

Consider this hypothetical: an old woman lives in public housing. She has worked her whole life in minimum-wage jobs to care for her grandchildren. Failed by the system, she's been denied the education with which she might protect herself, and falls easily prey to phishing emails targeting her lack of understanding of banking controls and IT norms. With digital surveillance tools, law enforcement is armed to catch the crooks who would defraud her.

Or this one: a young activist goes to Alabama to expose rampant voter suppression targeting minorities. The Republican party boss decides he's had enough, and tells his lieutenants he "wouldn't be sorry if that idiot wound up face down in the gutter." Thugs corner the activist late at night. Nobody is around to help. But the video camera police have installed on a streetlight catches it all. Even if police don't get there in time to stop the attack, they have the evidence they need to catch and prosecute the thugs who did it, to make sure they never do it again.

Historically, arms have been the tools of unjust oppression, but also of righteous revolution. They have been used to threaten, steal, and kill, but also to protect and secure. They are the tools of UN peacekeepers. They were in the hands of the National Guardsmen who forcibly desegregated the University of Alabama in 1963. Make no mistake, George Wallace was a monster, and he absolutely would have used violence on African American students had guns not been present to force him down. Surveillance is no different. It is a tool we can turn to good.

It pains me to see faith in law enforcement so badly eroded, even more to know that that erosion is so richly deserved. But the answer is not to throw the baby out with the bathwater. Surveillance is a tool that can be put to work for the people it has been used against. It can help, rather than harm.

Do not reject it. Reform it. Let it right itself and give our communities the protection they deserve.

CLOSING
▷ Con: Hamid Khan
& Ken Montenegro

DURING THIS DEBATE, WE have demonstrated that police surveillance is fundamentally flawed by design and is a tool of white supremacy. It is ironic that the other side acknowledges white supremacy and racism, yet reduces them to a momentary lapse of reason or function. White supremacy has been and remains essential to the core of the United States' social, cultural, economic, political, and structural DNA: it's not an action, it's the lived experience of millions upon millions who have been enslaved, genocided, incarcerated, deported, murdered. Police have been and remain the primary enforcement mechanism of white supremacy, and surveil-lance is the primary policing tool to trace, track, monitor, stalk, and murder.

In a rush to say surveillance is neutral, advocates of the Stalker State erase and minimize not only the experiences of marginalized and targeted com-munities, but also the sordid history of law enforcement surveillance (slave catching, union busting, COINTELPRO, etc).

The voices and history erased are exactly those of the people surveilled and harmed through narratives such as the "savage native," the "criminal black," the "illegal Latino," the "manipulative Asian," and the "terrorist Muslim." The history erased is the origin story of police as protectors of capital and of slave catchers. You cannot say you are interested in protecting marginalized communities while you actively criminalize them. You cannot admit that it is a tool of white supremacy yet assert that surveillance protects marginalized communities, while facts show us that police have not protected and do not protect these communities.

Police surveillance has pathologized black, brown, and poor folks as inher-ent risks to public safety. The very practice of police surveillance is based upon the need to lay bare, stalk, and harm the lives of "suspect" bodies and those deemed a threat to the system. Policing has always been and remains a primary tool for social control. Their violent method of control is to beat people into submission, throw them into cages, or murder them. This is white supremacy's way of dealing with the conditions of poverty, neglect, and racism that cause violence to occur in marginalized communities.

Surveillance is unleashed to decontextualize and criminalize targeted communities. Policing remains an instrument of oppression, surveilling populations while at the same time using force and coercion and reproducing fear within communities. Decontextualizing the history of surveillance dis-misses this key point: surveillance is a form of racialized violence that must be abolished, not reformed.

In the end, the history of police surveillance is a history of law enforce-ment leveraging surveillance and stalking for its intent to cause harm, with an increasing reliance on pseudoscience wrapped in the language of predictive algorithms, behavioral surveillance, and data mining (to name a few). If we were to adapt reform-based approaches, we would essentially be saying that we're looking for a reformed or kinder, gentler racism. Reforms and other tinkering around the edges of this violence create more harm than they remedy.

We invite people of conscience to work towards the abolition of law enforcement surveillance before the next bullet hits a body. Say no to complicity, say no to fear, say no to the Stalker State. ●

▷

A COMPENDIUM OF LAW ENFORCEMENT SURVEILLANCE TOOLS

By Edward F. Loomis

○ Facial Recognition Systems

● **GPS Trackers**

○ License Plate Readers

○ Drones

○ Body Cameras

○ Cell Tower Simulators

○ Parallel Construction

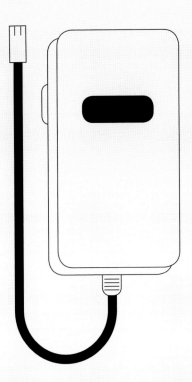

A GPS tracker is a device attached to or carried by a moving vehicle or person that receives Global Positioning System signals, allowing the device to track its movements and determine the location of the carrier at intervals. There are three varieties of GPS trackers: logger, beacon, and transponder.

Within its internal storage, a GPS logger system records a device's location—and the time it was recorded—based on the receipt of signals from the U.S. government's GPS satellites. The recordings are collected at set intervals and can later be downloaded to a computer for analysis.

The GPS beacon system also measures the device's location as determined by GPS satellites at regular intervals. The difference is that it

▷ **A Compendium of Law Enforcement Surveillance Tools**

○ Facial Recognition Systems
● GPS Trackers
○ License Plate Readers
○ Drones
○ Body Cameras
○ Cell Tower Simulators
○ Parallel Construction

MCS54 0092

transmits this information to a remote server instead of recording it in its internal storage, making its location available for instantaneous tracking.

A GPS transponder system is similar to a GPS beacon system in that it transmits the device's location and time to a remote server, but only in response to a command for the information by said server.

The privacy issues involving law enforcement use of GPS trackers become apparent in cases where police officers employ them without valid warrants. On November 8, 2011, a landmark GPS tracker case, *United States v. Jones*, was argued before the Supreme Court. In 2004, FBI and Washington Metropolitan Police initiated an investigation of two suspected cocaine traffickers based on information obtained from informants and police surveillance operations. In the course of the investigation, the police obtained a warrant to install and use a GPS logger in Washington, DC, on a Jeep belonging to Antoine Jones for a period of ten days. However, the tracker was not installed in Washington, but rather in Maryland, a day after the warrant had expired. It was left in place recording the vehicle's movements for twenty-eight days.

Based on reconstruction of the vehicle's travels during those twenty-eight days and correlation of the stops with Jones's cell phone records, an arrest of Antoine Jones and Lawrence Maynard occurred on October 24, 2005, along with searches and seizure of ninety-seven kilos of cocaine, almost one kilogram of crack cocaine, and $850,000 in cash. Jones and Maynard were tried in the U.S. District Court of DC in January 2008 and sentenced to life in prison for conspiracy to distribute cocaine and possession with the intent to distribute cocaine and cocaine base.

Jones appealed the admission of GPS evidence to the U.S. Court of Appeals, and in 2010 that Court ruled that Jones had a reasonable expectation of privacy in the movement of his vehicle on public streets. The Court also ruled that GPS surveillance of Jones constituted a search that violated the Fourth Amendment, thus requiring a valid search warrant, and reversed the lower Court ruling against Jones. The Court declared, "Without the GPS data the evidence that Jones was actually involved in the conspiracy is so far from 'overwhelming' that we are constrained to hold the Government has not carried its burden of showing the error was harmless beyond a reasonable doubt."

The Supreme Court accepted a Department of Justice petition to review the appellate court reversal on June 27, 2011. The Supreme Court concluded its review seven months later, issuing three separate opinions on the *United States v. Jones* case, each affirming the appellate court ruling and thus ensuring that Jones was cleared of all charges.

Currently no federal statutes exist on the topic of geolocation privacy or the question of how law enforcement, education institutions, or businesses can employ personal location information. However, several U.S. states have independently put in place

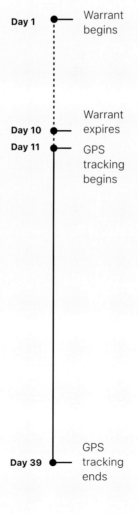

▽ **A TIMELINE OF THE UNWARRANTED SURVEILLANCE OF ANTOINE JONES**

Day 1 — Warrant begins

Day 10 — Warrant expires
Day 11 — GPS tracking begins

Day 39 — GPS tracking ends

personal location privacy laws, despite the fact that geolocation statutes at the federal level await action by congressional committees. A bill named the Geolocational Privacy and Surveillance Act (S. 395) was introduced in February 2017 and referred on the same day to the Senate Committee on the Judiciary. Its House equivalent (H.R. 3470) was introduced in July 2017 and referred to the House Subcommittee on Crime, Terrorism, Homeland Security, and Investigations a little over a month later. No further activity has since taken place in either chamber on the bill commonly known as the GPS Act. ●

Facebook headquarters, Menlo Park, CA.

The Economics of Mistrust

Ethan Zuckerman

Facebook has a trust problem. And so do we all.

On Sunday, March 25, 2018, the massive social-media company took out full-page ads in newspapers in the United States and United Kingdom. The ads apologized for missteps Facebook had made in releasing user data to Cambridge Analytica, the shady political consultancy that worked for the Trump 2016 campaign and in support of the UK Brexit vote.

Zuckerberg's apology starts out well: "We have a responsibility to protect your information. If we can't, we don't deserve it." But it rapidly founders as Zuckerberg talks about issues of trust. "You may have heard of a quiz app built by a university researcher who leaked Facebook data of millions of people in 2014. That was a breach of trust, and I'm sorry that we didn't do more at the time."

That's a carefully crafted pair of sentences. The personality quiz he's referring to was developed by the researcher Dr. Aleksandr Kogan, who was able to access data on tens of millions of Facebook users through the app. In interviews, Zuckerberg is clear that he sees the breach of trust as Kogan's sale of that data to Cambridge Analytica so it could be used to construct personality profiles of American voters.

Kogan's act was clearly unethical, even though Cambridge Analytica's personality profiles are more snake oil than the psychological superweapon Steve Bannon hoped they would be. But Zuckerberg's shifting of blame onto Kogan is disingenuous. The data Kogan collected was data Facebook had been providing to all app developers for nearly five years. Whether you were an academic researcher with hidden motives, an ambitious marketer, or someone like Ian Bogost, who created the surprisingly popular viral game Cow Clicker, designed as a satire of online games, you were rewarded with a deep cache of information not just about the people who chose to use your app but about their friends as well. It's this design decision that should cause us to question the trust we're putting in Zuckerberg and his compatriots.

Of course, this is bigger than Facebook. The overwhelming majority of content and services on the internet are supported by a single model: targeted advertising based on behavioral and psychographic data. This model has one notable benefit. It allows online providers to gain large audiences by offering their wares for free. But there are massive downsides. Websites are at war with themselves, hoping to keep their readers on their sites, while also needing to divert them onto advertisers' sites to make money. The incentive to collect any conceivable datum about a user or her behavior is the hope to gain an edge in targeting ads to her. Buying and selling this behavioral data becomes part of the revenue model of these sites. In the process, users get used to the idea that they are continuously under surveillance.

A world where we're constantly watched is one we learn, moment by moment, to mistrust. We know Facebook is exploiting us, we know our attention is being peddled to the highest bidder, and we know that even opting out of the system isn't an escape, as Facebook maintains profiles even on non-users. We know that Facebook will fail us again, releasing data to people we would never personally choose to trust. The revelations of Facebook's weaknesses come less as a shock to us than as part of the disappointing reality we've all become accustomed to, where we're forced to trust institutions that rarely deserve or reward the faith we put in them.

A logical response to this wave of disappointment would be to distrust all aggregators of data, all large, centralized systems, and move towards a world in which we trust no one. That vision, compelling on its surface, is harder to achieve than it looks.

The internet was built in a way that makes it possible to participate in discussions online without placing significant trust in third parties, so long as you're willing to do a little work. Consider email. I use Gmail to send, receive, and read my email, which means I trust the company with some of the most intimate details of my personal and professional life. Google has access to what my employer pays me, when my boss yells at me, and the details of my divorce.

You may not be willing to make the privacy compromises I've made. Fortunately, you can set up your own email system. First, simply register a domain name and set up appropriate DNS and MX records to point to the

Linux box you've just configured. Now download and install Postfix so you've got a functioning sendmail server, then pick an IMAP server to install and configure. You'll also need to set up a mail client to work with that IMAP server. Be sure and keep all these tools patched and up to date, as well as regularly downloading blacklists so you can attempt to keep your mailbox spam-free.

This will take you the better part of a day if you're a seasoned Unix system administrator. If you're not, you'll be well on your way to being a sysadmin by the time you're done. This is probably why when you search for "how to run your own email server," the second result is an article titled "Why You Shouldn't Try to Host Your Own Email."

Oh, and even if you do all this, Google will still get a copy of the message you've sent to me, because I decided to trust Gmail. So if you really want to keep your unencrypted correspondence out of the hands of snooping companies, you need to convince your friends to take the steps you've just taken.

(You're probably better off just using Signal, an app that uses powerful encryption to protect your chats and calls. But of course you're going to need to ensure that Open Whisper Systems, which makes the app, has no back doors in its code that gives their employees access to your mail. Speaking of code, you read all the code for Postfix and your IMAP server to ensure there were no back doors in that hand-rolled email system of yours, right? You *do* care about privacy, right?)

For most people–including many technologically sophisticated people– trusting Google turns out to make more sense than doing the work of managing our own communication tools. Unsurprisingly, Google's sysadmins are better at security than most of us are. What's true for email is at least as true for social media. It's possible to manage your own alternative to Facebook, or at least to Twitter–Mastodon offers a full-featured alternative to the character-constrained messaging service–but it requires time, know-how, and the cooperation of the people you want to communicate with. The alternative is convenience and ease, at the cost of putting your trust into a company that's unlikely to respond when you complain.

The trust we put in Facebook, Google, and other internet giants can–and likely will–be violated, either intentionally, as Facebook has done by sharing sensitive data with app-building partners, or unintentionally, in a data breach. While these companies talk about data as a precious asset like oil, it behaves more like toxic waste when it leaks or spills.

Even when companies work to protect our data and use it ethically, trusting a platform gives that institution control over your speech. Whoever is in control of these platforms, or in the center of a communications network, can control what you say and whom you say it to. Platforms routinely block speech they see as against their terms of service, because it either encourages hate or violence, is degrading or harassing, or otherwise violates community norms. Sometimes this censorship is visible, and users are blocked from posting certain speech that's in violation of service terms; other times it can be subtler, when speech is simply "de-prioritized," buried deep in the feeds shared with other users. Facebook's news feed, which favors stories the algorithms think will interest us, chooses hundreds of stories a day that we will likely never see, deselected for reasons we aren't allowed to know. We probably both want and need some sort of moderation to allow healthy communities to operate, as many online spaces descend into hateful anarchy without careful tending. But trusting a platform means trusting it to decide when you can't speak.

Ultimately, trust leads to centralization. It's difficult to leave Facebook precisely because so many people have decided to trust Zuckerberg and his company with their data. Platforms like Facebook benefit from network effects: they grow increasingly valuable as more people use them, because you can connect with more people. And because companies like Google and Facebook aren't shy about swallowing their competitors (and because U.S. regulators are quite shy about intervening), the companies we most trust can quickly become near-monopolies whom we're forced to trust, if only because they've eliminated their most effective competitors.

One response to systems we don't and shouldn't trust is to build a new system that obviates trust. Thousands of people and billions of dollars are now participating in an experiment in a "trustless" system: the blockchain. As Satoshi Nakamoto, the as yet unidentified and probably pseudonymous creator of Bitcoin and the blockchain, explained in an early paper, "The root problem with conventional currency is all the trust that's required to make it work. The central bank must be trusted not to debase the currency, but the history of fiat currencies is full of breaches of that trust." Nakamoto was worried about trusting not just central banks but also commercial banks and transaction providers like Visa: "Commerce on the internet has come to rely almost exclusively on financial institutions serving as trusted third parties to process electronic payments. While the system works well enough for most transactions, it still suffers from the inherent weaknesses of the trust-based model."

Using some very clever mathematics, Nakamoto built a system that eliminated the requirement for a trusted third party to ensure that digital currencies aren't "double spent." Unlike a physical token (a coin), a bitcoin is nothing but a string of numbers, and the only thing that prevents you from spending it multiple times (and cheating the system) is someone watching to ensure that when Alice gives Bob a coin, her account is debited and his is credited. Instead of asking Visa, a bank, or some other third party to manage these transactions, Nakamoto created a distributed ledger, a publicly reviewable table of all transactions. This table is verified by thousands of computers, owned by different groups, that compete to verify the work as quickly as possible, as the winner receives a newly minted bitcoin for their efforts.

Bitcoin works, sort of. It was initially proposed as a method for conducting micro-transactions online, payments of a few cents or less, which aren't economically feasible to conduct over systems like Visa. But the system has proven too slow and too costly to make such transactions possible at present. Advocates then hoped that it would work as a "store of value," an asset that can be stored and retrieved at a later date with an expectation that it will continue to be valuable. But Bitcoin has proven to be vastly more volatile than traditional stores of value like gold, leading to people buying too high and losing their shirts. The novelty of the market means it's filled with naive investors and currency speculators–easy prey for thieves, who've succeeded in hacking the exchanges where bitcoins are traded and siphoning off millions of dollars in real money.

The real way in which Bitcoin–along with all contemporary blockchains–is flawed is that it's unbelievably expensive in terms of electricity and computer time. This is by design. Bitcoin relies on "proof of work" to validate existing transactions–essentially, thousands of computers have to execute millions of attempts to solve a math problem before anyone else does to verify a set of transactions. If it were easy to solve these problems, it would be simple to inject false information into the public ledger, giving yourself coins you're not entitled to. To avoid the problem of trusting a central entity, Bitcoin makes it prohibitively expensive to commit fraud.

The net result is that Bitcoin is orders of magnitude less efficient than a centralized system like Visa. The giant farms of computers that validate the distributed ledger and mine bitcoins in the process are now estimated to consume 0.27 percent of the world's electricity, approximately the same percentage as the nation of Colombia uses. (This figure changes daily, and will likely be higher by the time you read this–the Bitcoin Energy Consumption Index has current estimates.) Bitcoin boosters will tell you either that this cost will go down once a new "proof of stake" algorithm replaces the costly proof of work systems now in place, or that the expense is worth it to have a trust-free currency. The first may eventually prove to be true, while the second is essentially a statement of religious belief.

All this is to say mistrust is inherently costly. Francis Fukuyama detailed this in his book *Trust*, where he examined the economics of high- and low-trust communities in Italy. Fukuyama discovered that in communities where trust in strangers was low, it was very difficult for firms to grow and expand. The reasons are fairly simple: consider the challenges of doing business with someone you've never met before versus doing business with a longtime partner. Because you trust your partner, you're willing to ship her goods before her check clears because you trust that she'll make you whole, speeding the entire process. Transacting with a stranger involves multiple, cautious, and ultimately costly steps where both sides verify the other party's good faith before moving forward.

Transacting in a world without trust is a lot like doing business in the early days of eBay. As people began selling each other PEZ dispensers and PCs online, they discovered a plethora of ways to defraud each other online. So eBay was forced to build a reputation system, tracking participant reports of who was reliable. Then they built an escrow system, providing assurance by acting as a trusted third party and ensuring that expensive items were delivered before releasing payments. They acquired and then spent vast annual sums on PayPal so they could design a currency that would resist some of the more obvious forms of online fraud. You can think of much of eBay's multibillion-dollar market capitalization as the value of providing trust to an otherwise trust-free marketplace.

It's possible that mistrust is similarly expensive in our political lives, a worrisome concept at a moment where trust in government and social institutions is hitting historic lows. In 1964, 77 percent of Americans reported that they trusted the government to do the right thing all or most of the time. That figure is now 18 percent. While the government, and specifically Congress, is subject to very low levels of trust in the United States, virtually all institutions—churches, banks, the press, the medical system, the police—are trusted significantly less than they were in the 1970s and 1980s.

As a result, it's difficult for institutions to exercise the power they once had. When Americans look nostalgically to a period of post-WWII prosperity and growth, they are looking back to a moment when people trusted the government to build highways and bridges, to support college educations and mortgages, and to use the powers of taxation and spending to build public goods and reduce inequality. The trust in government that allowed both the interstate highway system and the success of the civil rights movement has been replaced with a pervasive skepticism of all large institutions.

Donald Trump got elected by harnessing this mistrust. Promising to increase trustworthiness by "draining the swamp," he's instead governed by actively seeking to erode the electorate's trust in institutions. His mantra of "fake news" is designed not only to drive supporters towards Trump-friendly media sources like Fox News and Sinclair, but also to diminish our trust in the ability of independent actors to review or check his power. At least as dangerous is the "deep state" narrative raised in corners of the Trumposphere and amplified by the President's attacks on the FBI and other government institutions. In positing a conspiracy against the presidency led by partisan holdovers of previous administrations, Trump invites his supporters to imagine a world where anyone who speaks against him is a coup plotter, attempting to undermine his power for sinister partisan purposes.

We know that the corrosion of trust in institutions makes it harder for governments to build infrastructures that link people together and encourage economic growth. We know we're losing trust in different institutions–the church, the government–that encouraged us to work together towards common goals. What we don't know is what this perpetual low-level mistrust does to us as individuals.

I got an illustrative example the other day, when a woman wrote to me worried that the U.S. Postal Service was censoring her mail. (Write enough about privacy and surveillance and you, too, will receive these emails.) The books she orders about organizing and civil rights never arrive, which means that the post office is monitoring her correspondence and preventing her from receiving subversive texts ordered from Amazon.

I spent a while composing a response to this woman. It began with the idea that the American experiment in democracy was quite literally based around the post office, which represented more than three-quarters of the jobs in the federal government in 1831. I wrote about how Benjamin Franklin imagined a republic of letters, in which a public sphere of freely exchanged newspapers allowed the citizens of a new nation to debate ideas and govern themselves. I outlined the laws that prevent the postal service from interfering with the mail, the postal inspectors who are responsible for enforcing those laws, the congressional oversight of the postal service, and the press stories written about the reports generated by those congressional hearings.

While I was writing, she wrote back to tell me she'd received her book.

Our Eroding Confidence in U.S. Institutions

CHURCH OR ORGANIZED RELIGION

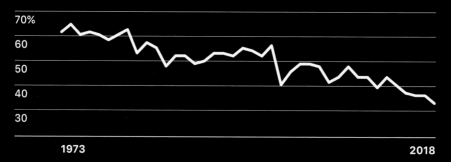

1973   2018

SUPREME COURT

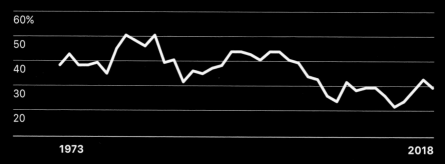

1973   2018

CONGRESS

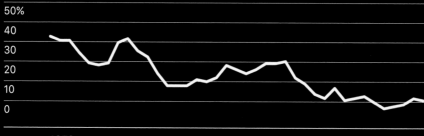

1973   2018

NEWSPAPERS　　　　% OF PEOPLE WHO REPORTED STRONG CONFIDENCE

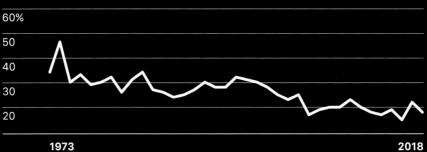

BANKS

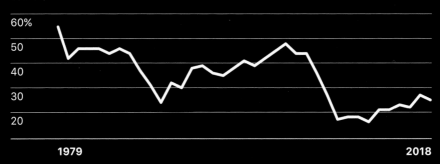

TV NEWS

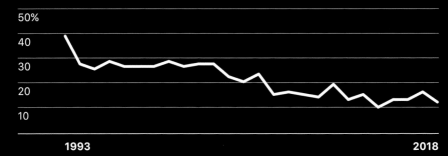

Source: Gallup

"So maybe it didn't happen this time. But how could you really know if they wanted to keep information from reaching you?"

That's a surprisingly deep question. If you trust institutions–the post office, Congress, the press–there's a predictable system of oversight that includes actors inside and outside the system, backed by a long history of norms that strongly discourage the sorts of behavior she is worried about. But if you don't trust institutions–say, if the most visible face of the government is on TV every day telling you to fear a deep-state conspiracy and not to trust the lying media–how would you know if that missing book was a shipping error or something more sinister? Once we've lost trust in the institutions that are designed to protect and oversee the systems we depend on, are we responsible for auditing those institutions ourselves? Or must we simply live with a perpetual low-grade mistrust in, well, everything?

There's no way of verifying what Facebook is doing with our data. In 2011, well before the revelations about Cambridge Analytica, Facebook had its wrist slapped by the Federal Trade Commission for telling users they could keep their data private, and making it accessible to third parties. Facebook agreed to a set of changes–a consent decree–certifying that it would behave better in the future. Cambridge Analytica–which we know of only because a whistleblower came forth–suggests little reason to trust Facebook with our privacy or the FTC with oversight.

Digital-security advocates diagnose a dangerous condition called "privacy nihilism." It occurs when you understand just how many ways your information and identity can be compromised online and simply give up on protecting yourself. It's a stance that can seem deeply knowledgeable and worldly, but leaves you more vulnerable to exploitation than if you had taken imperfect but wise, defensive steps. The weaponization of mistrust in American politics suggests a deeper form of nihilism, a realization that our political systems are so fragile, so dependent on a set of norms preventing outrageous behavior, so subject to capture by parties that favor partisanship over compromise, that checks and balances can quickly become rubber stamps.

The solution to mistrust in online systems isn't to eliminate trust entirely, as the blockchain seeks to do. It's to build systems that help us trust better and more wisely–eBay's reputation systems have provided sufficient enough assurance that selling our old junk online is now as routine as holding a garage sale. Airbnb is built on the strange notion that we can trust a stranger to rent us a room for the night, or to welcome a stranger into our spare bedroom. When we climb into an Uber to head to the restaurant recommended by Yelp, we are not living in a trust-free world, but one where we've learned to trust deeply in strangers.

These systems are far from perfect. But they work well enough, most of the time, that they generate enormous value for those who use them. At a moment where many systems, from health care to Congress, seem to be failing us, the question may not be how we restore trust in broken systems, but how we design new systems that let us cooperate and trust each other.

Facebook may not deserve our trust, nor may the United States government. But to respond with nihilism is to exit the playing field and cede the game to those who exploit mistrust. We need to harness our mistrust, to use it as fuel. This means forsaking the expensive fantasy of a trustless world and doing the hard work of building new systems that deserve our trust. ●

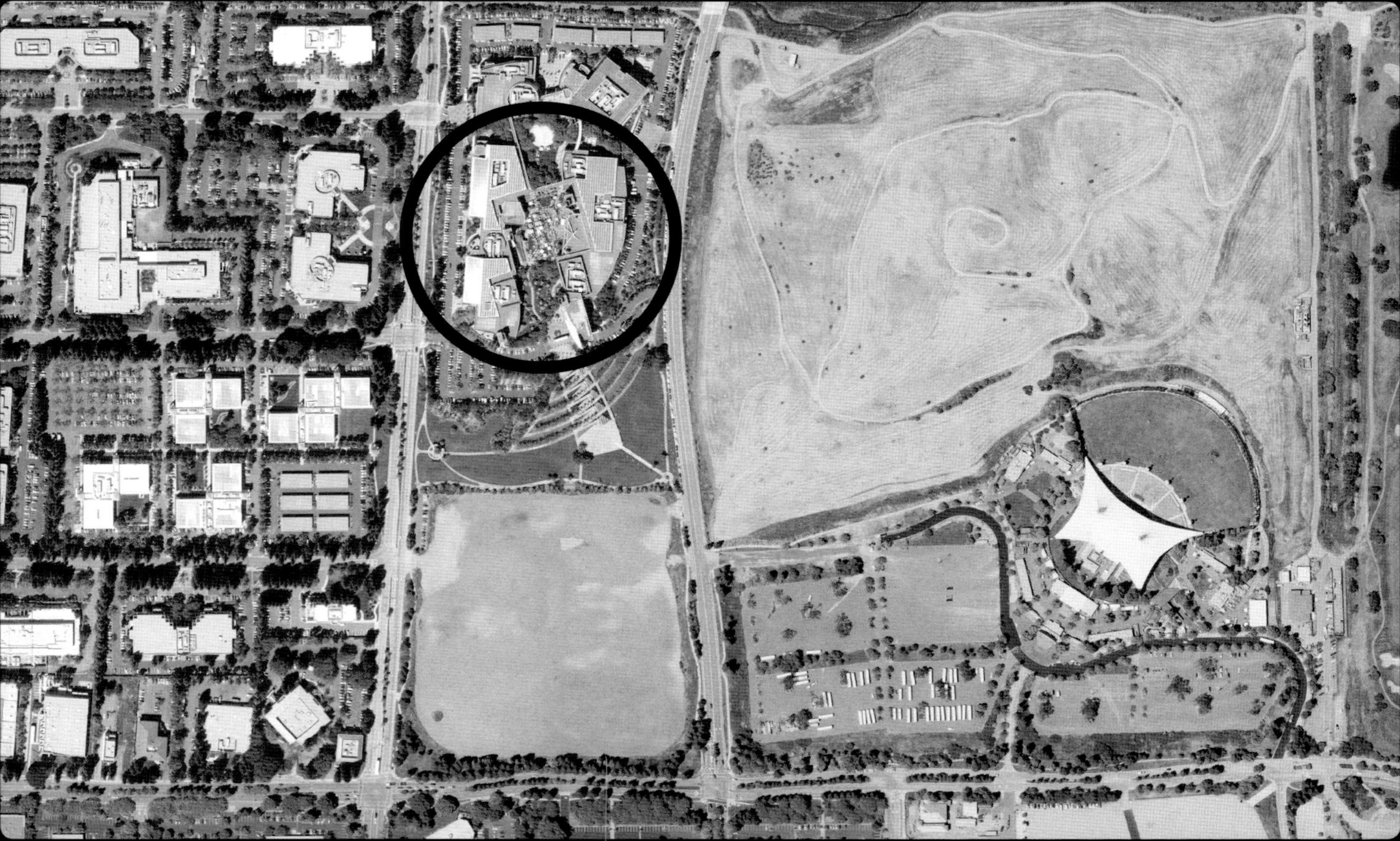

Google headquarters, Mountain View, CA.

Photo: USGS

Search Queries of Visitors Who Landed at the Online Litmag *The Big Ugly Review,* but Who, We Are Pretty Sure, Were Looking for Something Else

Elizabeth Stix

ow many times have you asked yourself, "How much meat is in my intestines?" Or "Will a temp of 5 kill a cat?" Maybe your mind turned to darker things: "Does an ugly ultrasound mean an ugly baby?" or "Does heavy bleeding ever stop?"

Who can answer these questions that itch at our minds? A trusted friend, perhaps. Or a paramedic. Some perils–"Daddy is dying and won't speak to me"–are best left to a therapist, or a psychic. Or God.

Or, as it turns out, me.

One afternoon, while I was obsessing over Google Analytics, I discovered a strange window to the human soul. I publish an online journal called *The Big Ugly Review*, which showcases fiction, essays, photography, short films, and music. When our web designer installed Google Analytics, I became intrigued. What were people looking for when they came to the site?

Google Analytics allows you to track almost everything about who is visiting your website–where they live, what browser they use, what they looked at, how long they stayed. And if they came from Google, it tells you the search query that brought them there.

What I found surprised me: a huge number of readers stumbled upon the magazine because of a coincidental overlap between a term in their search query and a word or phrase on our site. Because of the many unusual story titles (from "My Eczema, Myself" and "Suburban Hottentot" to "Six Things I Will Not Say Tomorrow at My Father's Funeral"), I found that I had become an inadvertent portal to thousands of bizarre searches.

These queries revealed much about the way people turn to the internet, and to Google in particular. It bared an unexpected human frailty that no searcher realized would ever be captured and analyzed by a stranger. These people were not just searching for answers. It was almost as if they thought they were talking to God.

They shared their most poignant fears and insecurities. They were turning to Google the way people used to kneel down before an oracle, humble and beseeching. To a generation that can type any question, any time, with the illusion of privacy, perhaps Google *is* the new God.

For some people, all they need is quick advice. They have things they just want to know.

> I grew a flap on my asshole what is it?
> What to say to the principal when you called someone big tits
> When a guy gives you a wink goodbye instead of a hug
> Do women like getting facials during sex?

Others whisper their confessions:

> Am I a needy wife?
> Am I a bad person for wanting a bigger diamond?
> I regret letting wife leave

Others aim high–perhaps too high to ask the question out loud:

> How to make a girl like you even though she
> doesn't like you and your really ugly
>
> Woman who are after sex, who are ugly, who are local

Sometimes they put parts of their question in quotation marks. It's that phrase that they definitely want in there, in exactly that order.

> "things i would not do" if i was in star wars
> oh "my god" "did you hear" how big he is
> "headphones" "roommate" "masterbation"

Some queries are so strange that even if Google *were* God, would God know the answer?

> Ark, me, dating
> My boyfriend walked in my room and saw our baby's head
> In plane couch mom spreads feets

Sometimes, when you read them all together, they begin to sound like poetry:

She, A Love Story

She kissed my thigh
She said "his penis"
She scratched long nails
She sit on big engine throbbing
She spanked me over her knee
She gently diapered me
She never came home.

These people are in pain. They just want to tell someone. But the wise, confidential friend they share it with isn't sitting beside them with a cup of tea. Although they don't know it... it's me.

I feel ugly with my rubber bands and braces

I have a big ugly nose no sex date

My big ass is in the way

I think about my self that I am ugly

I want to know about the utrus

Feel ugly walk around in cute outfits boyfriend wont touch me

I will cry tomorrow

But I'm not afrais, I'll be brave. I couldn't back anything
that I've gone through.

I will be changed by the things that have happened to me,
but I refuse to be reduced by them.

I am glad they can't see me. I'm glad I'm not on the hook for this, as much as they are suffering. What would I tell this person if I could?

alcohol yellow eyes red skin shaking head how long left

Or this one?

> how should i feel after my mom starts dating
> again after my dad dies?

What would you advise here?

> Woman or wife or girl or friend or cut or severed or
> chopped or lopped or sliced or bit his

I generated a report of the search queries and it ran to over seven hundred pages, more than twenty thousand search terms. Here I sit, eavesdropping on people's most shameful anxieties, their deepest fears, their most perverted fantasies. But of course, it's not only me. In this age of internet hacks, surveillance, and data sharing, people must know that everything they type into the computer is fair game. It's not even some nefarious agent who will reveal all our secrets in the end. Got genital warts? Go ahead and look it up, but the next time you are scrolling through Facebook, know that every banner ad will be for wart removal.

Back in the day when most people who confessed sins did so from behind a discreet wooden panel, there was some plausible deniability that the listener didn't know who was on the other side. As we whisper into our keyboards, we cling to that false security still. It's all there, though–all our worries, anxieties, secrets–stored in the servers at Google, and Yahoo, and every company with whom we have knowingly or unknowingly signed a user agreement. All quietly waiting to come back and haunt us.

Or maybe God does lie in between the zeros and ones of the internet. Maybe these searchers found what they were looking for. Perhaps fate brought them to my magazine, and somewhere between the stories about the first dog sent into space and streets paved with candy, their prayers were answered. I hope so. ●

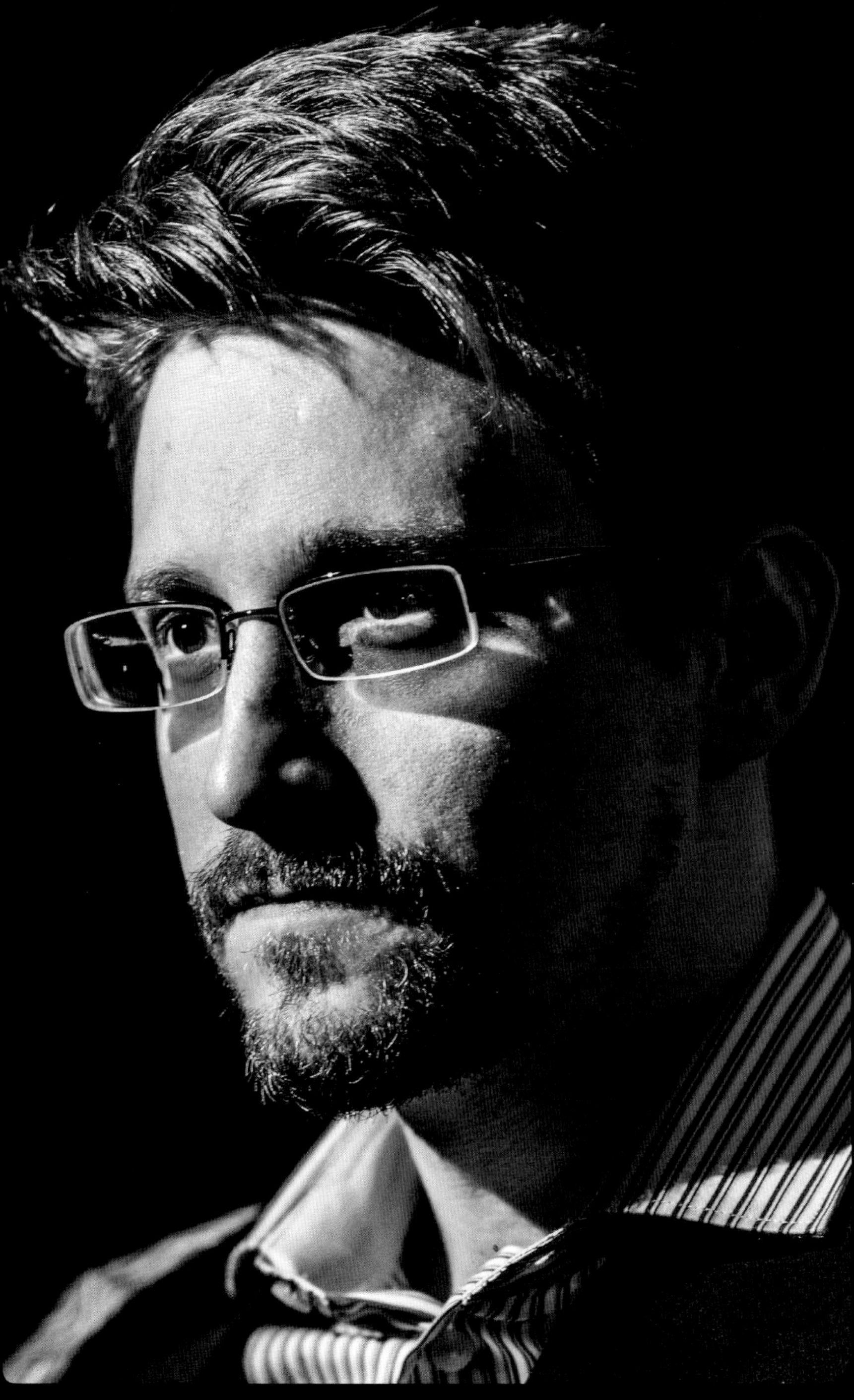
Photo: Lindsay Mills

The NSA whistleblower and his math-challenged lawyer discuss the nuts and bolts of blockchain and how it could alter the internet, and trust itself.

Edward Snowden Explains Blockchain to His Lawyer— and the Rest of Us

Ben Wizner

O ver the last five years, Edward Snowden and I have carried on an almost daily conversation, most of it unrelated to his legal troubles. Sometimes we meet in person in Moscow over vodka (me) and milkshakes (him). But our friendship has mostly taken place on secure messaging platforms, a channel that was comfortable and intuitive for him but took some getting used to for me. I learned to type with two thumbs as we discussed politics, law, and literature; family, friends, and foster dogs. Our sensibilities are similar but our worldviews quite different: I sometimes accuse him of technological solutionism; he accuses me of timid incrementalism.

Through it all, I've found him to be the clearest, most patient, and least condescending explainer of technology I've ever met. I've often thought that I wished more people–or perhaps different people–could eavesdrop on our conversations. What follows is a very lightly edited transcript of one of our chats. In it, Ed attempts to explain "blockchain" to me, despite my best efforts to cling to my own ignorance.

BEN WIZNER: The Electronic Frontier Foundation recently joked that "the amount of energy required to download tweets, articles, and instant messages which describe what 'the blockchain' is and how 'decentralized' currencies are 'the future' will soon eclipse the total amount of power used by the country of Denmark." It's true that there are a lot of "blockchain explainers" out there. And yet I'm ashamed to admit I still don't really get it.

EDWARD SNOWDEN: Are you asking for another math lesson? I've been waiting for this day. You remember what a cryptographic hash function is, right?

BW: This is where I'm supposed to make a joke about drugs. But no, I do not now nor will I ever remember that.

ES: Challenge accepted. Let's start simpler: what do you know about these mythical blockchains?

BW: That I could have been rich if I'd listened to you about this four years ago? But really, I've heard a lot and understood little. "Decentralized." "Ledgers." What the hell is a blockchain?

ES: It's basically just a new kind of database. Imagine updates are always added to the end of it instead of messing with the old, preexisting entries–just as you could add new links to an old chain to make it longer–and you're on the right track. Start with that concept, and we'll fill in the details as we go.

BW: Okay, but why? What is the question for which blockchain is the answer?

ES: In a word: trust. Imagine an old database where any entry can be changed just by typing over it and clicking save. Now imagine that entry holds your bank balance. If somebody can just arbitrarily change your balance to zero, that kind of sucks, right? Unless you've got student loans.

The point is that any time a system lets somebody change the history with a keystroke, you have no choice but to trust a huge number of people to be both perfectly good and competent, and humanity doesn't have a great track record of that. Blockchains are an effort to create a history that can't be manipulated.

BW: A history of what?

ES: Transactions. In its oldest and best-known conception, we're talking about Bitcoin, a new form of money. But in the last few months, we've seen efforts to put together all kind of records in these histories. Anything that needs to be memorialized and immutable. Health-care records, for example, but also deeds and contracts.

When you think about it at its most basic technological level, a blockchain is just a fancy way of time-stamping things in a manner that you can prove to posterity hasn't been tampered with after the fact. The very first bitcoin ever created, the "Genesis Block," famously has one of those "general attestations" attached to it, which you can still view today.

It was a cypherpunk take on the old practice of taking a selfie with the day's newspaper, to prove this new bitcoin blockchain hadn't secretly been created months or years earlier (which would have let the creator give himself an unfair advantage in a kind of lottery we'll discuss later).

BW: Blockchains are a history of transactions. That's such a letdown. Because I've heard some extravagant claims like: blockchain is an answer to censorship. Blockchain is an answer to online platform monopolies.

ES: Some of that is hype cycle. Look, the reality is blockchains can theoretically be applied in many ways, but it's important to understand that mechanically, we're discussing a very, very simple concept, and therefore the applications are all variations on a single theme: verifiable accounting. Hot.

So, databases, remember? The concept is to bundle up little packets of data, and that can be anything. Transaction records, if we're talking about money, but just as easily blog posts, cat pictures, download links, or even moves in the world's most over-engineered game of chess. Then, we stamp these records in a complicated way that I'm happy to explain despite protest, but if you're afraid of math, you can think of this as the high-tech version of a public notary. Finally, we distribute these freshly notarized records to members of the network, who verify them and update their independent copies of this new history. The purpose of this last step is basically to ensure no one person or small group can fudge the numbers, because too many people have copies of the original.

It's this decentralization that some hope can provide a new lever to unseat today's status quo of censorship and entrenched monopolies. Imagine that instead of today's world, where publicly important data is often held exclusively at GenericCorp LLC, which can and does play God with it at the public's expense, it's in a thousand places with a hundred jurisdictions. There is no takedown mechanism or other "let's be evil" button, and creating one requires a global consensus of, generally, at least 51 percent of the network in support of changing the rules.

BW: So even if Peter Thiel won his case and got a court order that some article about his vampire diet had to be removed, there would be no way to enforce it. Yes? That is, if *Blockchain Magazine* republished it.

ES: Right—so long as *Blockchain Magazine* is publishing to a decentralized, public blockchain, they could have a judgment ordering them to set their office on fire and it wouldn't make a difference to the network.

BW: So... how does it work?

ES: Oh man, I was waiting for this. You're asking for the fun stuff. Are you ready for some abstract math?

BW: As ready as I'll ever be.

ES: Let's pretend you're allergic to finance, and start with the example of an imaginary blockchain of blog posts instead of going to the normal Bitcoin examples. The interesting mathematical property of blockchains, as mentioned earlier, is their general immutability a very short time past the point of initial publication.

For simplicity's sake, think of each new article published as representing a "block" extending this blockchain. Each time you push out a new article, you are adding another link to the chain itself. Even if it's a correction or update to an old article, it goes on the end of the chain, erasing nothing. If your chief concerns were manipulation or censorship, this means once it's up, it's up. It is practically impossible to remove an earlier block from the chain without also destroying every block that was created after that point and convincing everyone else in the network to agree that your alternate version of the history is the correct one.

Let's take a second and get into the reasons for why that's hard. So, blockchains are record-keeping backed by fancy math. Great. But what does that mean? What actually stops you from adding a new block somewhere other than the end of the chain? Or changing one of the links that's already there?

We need to be able to crystallize the things we're trying to account for: typically a record, a timestamp, and some sort of proof of authenticity.

So on the technical level, a blockchain works by taking the data of the new block–the next link in the chain–stamping it with the mathematic equivalent of a photograph of the block immediately preceding it and a timestamp (to establish chronological order of publication), then "hashing it all together" in a way that proves the block qualifies for addition to the chain.

BW: "Hashing" is a real verb?

ES: A cryptographic hash function is basically just a math problem that transforms any data you throw at it in a predictable way. Any time you feed a hash function a particular cat picture, you will always, always get the same number as the result. We call that result the "hash" of that picture, and feeding the cat picture into that math problem "hashing" the picture. The key concept to understand is that if you give the very same hash function a slightly different cat picture, or the same cat picture with even the tiniest modification, you will get a WILDLY different number ("hash") as the result.

BW: And you can throw any kind of data into a hash function? You can hash a blog post or a financial transaction or *Moby-Dick*?

ES: Right. So we hash these different blocks, which, if you recall, are just glorified database updates regarding financial transactions, web links, medical records, or whatever. Each new block added to the chain is identified and validated by its hash, which was produced from data that intentionally includes the hash of the block before it. This unbroken chain leads all the way back to the very first block, which is what gives it the name.

I'm sparing you some technical nuance here, but the important concepts to understand are that blocks in the chain are meant to be verifiable, strictly ordered by chronology, and immutable. Each new block created, which in

The Value of Bitcoin Over Time

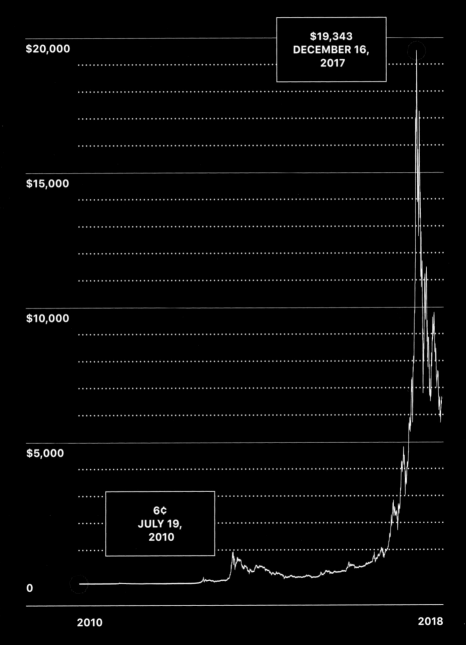

$20,000

$19,343
DECEMBER 16, 2017

$15,000

$10,000

$5,000

6¢
JULY 19, 2010

0

2010

2018

Source: CoinDesk

the case of Bitcoin happens every ten minutes, effectively testifies about the precise contents of all the ones that came before it, making older blocks harder and harder to change without breaking the chain completely.

So by the time our Peter Thiel catches wind of the story and decides to kill it, the chain has already built a thousand links of confirmable, published history.

BW: And this is going to... save the internet? Can you explain why some people think blockchain is a way to get around or replace huge tech platform monopolies? Like how could it weaken Amazon? Or Google?

ES: I think the answer there is "wishful thinking." At least for the foreseeable future. We can't talk Amazon without getting into currency, but I believe blockchains have a much better chance of disrupting trade than they do publication, due to their relative inefficiency.

Think about our first example of your bank balance in an old database. That kind of setup is fast, cheap, and easy, but makes you vulnerable to the failures or abuses of what engineers call a "trusted authority." Blockchains do away with the need for trusted authorities at the expense of efficiency. Right now, the old authorities like Visa and MasterCard can process tens of thousands of transactions a second, while Bitcoin can only handle about seven. But methods of compensating for that efficiency disadvantage are being worked on, and we'll see transaction rates for blockchains improve in the next few years to a point where they're no longer a core concern.

BW: I've been avoiding this, because I can't separate cryptocurrency from the image of a bunch of tech bros living in a palace in Puerto Rico as society crumbles. But it's time for you to explain how Bitcoin works.

ES: Well, I hate to be the bearer of bad news, but Zuckerberg is already rich.

Money is, of course, the best and most famous example of where blockchains have been proven to make sense.

BW: With money, what is the problem that blockchain solves?

ES: The same one it solves everywhere else: trust. Without getting too abstract: what *is* money today? A little cotton paper at best, right? But most of the time, it's just that entry in a database. Some bank says you've got three hundred rupees today, and you really hope they say the same or better tomorrow.

Now think about access to that reliable bank balance–that magical number floating in the database–as something that can't be taken for granted, but is instead transient. You're one of the world's unbanked people. Maybe you don't meet the requirements to have an account. Maybe banks are unreliable where you live, or, as happened in Cyprus not too long ago, they decided to seize people's savings to bail themselves out. Or maybe the money itself is unsound, as in Venezuela or Zimbabwe, and your balance from yesterday that could've bought a house isn't worth a cup of coffee today. Monetary systems fail.

BW: Hang on a minute. Why is a "bitcoin" worth anything? What generates value? What backs the currency? When I own a bitcoin, what do I really own?

ES: Good question. What makes a little piece of green paper worth anything? If you're not cynical enough to say "men with guns," which are the reason legal tender is treated different from Monopoly money, you're talking about scarcity and shared belief in the usefulness of the currency as a store of value or a means of exchange.

Let's step outside of paper currencies, which have no fundamental value, to a more difficult case: why is gold worth so much more than its limited but real practical uses in industry? Because people generally agree it's worth more than its practical value. That's really it. The social belief that it's expensive to dig out of the ground and put on a shelf, along with the expectation that others are also likely to value it, transforms a boring metal into the world's oldest store of value.

Blockchain-based cryptocurrencies like Bitcoin have very limited fundamental value: at most, it's a token that lets you save data into the blocks of their respective blockchains, forcing everybody participating in that blockchain to keep a copy of it for you. But the scarcity of at least some cryptocurrencies is very real: as of today, no more than twenty-one million bitcoins will ever be created, and seventeen million have already been claimed. Competition to "mine" the remaining few involves hundreds of millions of dollars' worth of equipment and electricity, which economists like to claim are what really "backs" Bitcoin.

Yet the hard truth is that the only thing that gives cryptocurrencies value is the belief of a large population in their usefulness as a means of exchange. That belief is how cryptocurrencies move enormous amounts of money across the world electronically, without the involvement of banks, every single day. One day capital-B Bitcoin will be gone, but as long as there are people out there who want to be able to move money without banks, cryptocurrencies are likely to be valued.

BW: But what about you? What do you like about it?

ES: I like Bitcoin transactions in that they are impartial. They can't really be stopped or reversed, without the explicit, voluntary participation by the people involved. Let's say Bank of America doesn't want to process a payment for someone like me. In the old financial system, they've got an enormous amount of clout, as do their peers, and can make that happen. If a teenager in Venezuela wants to get paid in a hard currency for a web development gig they did for someone in Paris, something prohibited by local currency controls, cryptocurrencies can make it possible. Bitcoin may not yet really be private money, but it is the first "free" money.

Bitcoin has competitors as well. One project, called Monero, tries to make transactions harder to track by playing a little shell game each time anybody spends money. A newer one by academics, called Zcash, uses novel math to enable truly private transactions. If we don't have private transactions by default within five years, it'll be because of law, not technology.

BW: So if Trump tried to cut off your livelihood by blocking banks from wiring your speaking fees, you could still get paid.

ES: And all he could do is tweet about it.

BW: The downside, I suppose, is that sometimes the ability of governments to track and block transactions is a social good. Taxes. Sanctions. Terrorist finance.

We want you to make a living. We also want sanctions against corrupt oligarchs to work.

ES: If you worry the rich can't dodge their taxes without Bitcoin, I'm afraid I have some bad news. Kidding aside, this is a good point, but I think most would agree we're far from the low-water mark of governmental power in the world today. And remember, people will generally have to convert their magic internet money into another currency in order to spend it on high-ticket items, so the government's days of real worry are far away.

BW: Explore that for me. Wouldn't the need to convert Bitcoin to cash also affect your Venezuelan teen?

ES: The difference is scale. When a Venezuelan teen wants to trade a month's wages in cryptocurrency for her local currency, she doesn't need an ID check and a bank for that. That's a level of cash people barter with every day, particularly in developing economies. But when a corrupt oligarch wants to commission a four hundred million-dollar pleasure yacht, well, yacht builders don't have that kind of liquidity, and the existence of invisible internet money doesn't mean cops won't ask how you paid for it.

The off-ramp for one is a hard requirement, but the other can opt for a footpath.

Similarly, it's easier for governments to work collectively against "real" criminals–think bin Laden–than it is for them to crack down on dissidents like Ai Weiwei. The French would work hand in hand with the Chinese to track the activity of bin Laden's Bitcoin wallet, but the same is hopefully not true of Ai Weiwei.

BW: So basically you're saying that this won't really help powerful bad actors all that much.

ES: It could actually hurt them, insofar as relying on blockchains will require them to commit evidence of their bad deeds onto computers, which, as we've learned in the last decade, government investigators are remarkably skilled at penetrating.

BW: How would you describe the downsides, if any?

ES: As with all new technologies, there will be disruption and there will be abuse. The question is whether, on balance, the impact is positive or negative. The biggest downside is inequality of opportunity: these are new technologies that are not that easy to use and still harder to understand. They presume access to a level of technology, infrastructure, and education that is not universally available. Think about the disruptive effect globalization has had on national economies all over the world. The winners have won by miles, not inches, with the losers harmed by the same degree. The first-mover advantage for institutional blockchain mastery will be similar.

BW: And the internet economy has shown that a platform can be decentralized while the money and power remain very centralized.

ES: Precisely. There are also more technical criticisms to be made here, beyond the scope of what we can reasonably get into. Suffice it to say cryptocurrencies are normally implemented today through one of two kinds of lottery systems, called "proof of work" and "proof of stake," which are a sort of necessary evil arising from how they secure their systems against attack. Neither is great. "Proof of work" rewards those who can afford the most infrastructure and consume the most energy, which is destructive and slants the game in favor of the rich. "Proof of stake" tries to cut out the environmental harm by just giving up and handing the rich the reward directly, and hoping their limitless, rent-seeking greed will keep the lights on. Needless to say, new models are needed.

BW: Say more about the environmental harms. Why does making magical internet money use so much energy?

ES: Okay, imagine you decide to get into "mining" bitcoins. You know there are a limited number of them up for grabs, but they're coming from somewhere, right? And it's true: new bitcoins will still continue to be created every ten minutes for the next couple years. In an attempt to hand them out fairly, the original creator of Bitcoin devised an extraordinarily clever scheme: a kind of global math contest. The winner of each roughly ten-minute round gets that round's reward: a little treasure chest of brand new, never-used bitcoins, created from the answer you came up with to that round's math problem. To keep all the coins in the lottery from being won too quickly, the difficulty of the next math problem is increased based on how quickly the last few were solved. This mechanism is the explanation of how the rounds are always roughly ten minutes long, no matter how many players enter the competition.

The flaw in all of this brilliance was the failure to account for Bitcoin becoming too successful. The reward for winning a round, once worth mere pennies, is now around one hundred thousand dollars, making it economically reasonable for people to divert enormous amounts of energy, and data centers full of computer equipment, toward the math–or "mining"–contest. Town-sized Godzillas of computation are being poured into this competition, ratcheting the difficulty of the problems beyond comprehension.

This means the biggest winners are those who can dedicate tens of millions of dollars to solving a never-ending series of problems with no meaning beyond mining bitcoins and making its blockchain harder to attack.

BW: "A never-ending series of problems with no meaning" sounds like… nihilism. Let's talk about the bigger picture. I wanted to understand blockchains because of the ceaseless hype. Some governments think that Bitcoin is an

Bitcoin Energy Consumption

ORDOS
•
Bitmain data center

$39K

▲ Daily electricity bill of the Bitmain data center in Inner Mongolia, China.

6.6M

▲ Number of U.S. households that could be powered by Bitcoin. This amounts to all households in the four largest cities of the U.S., plus those of Wyoming.*

NEW YORK CITY   HOUSTON   CHICAGO   LOS ANGELES   WYOMING

*There is no clear consensus as to the best way of estimating Bitcoin's energy consumption, and this estimate is subject to shift over time. The above statistics are based on the Bitcoin Energy Consumption Index from July 2018.

Source: Digiconomist

existential threat to the world order, and some venture-capital types swear that blockchains will usher in a golden age of transparency. But you're telling me it's basically a fancy database.

ES: The tech is the tech, and it's basic. It's the applications that matter. The real question is not "what is a blockchain," but "how can it be used?" And that gets back to what we started on: trust. We live in a world where everyone is lying about everything, with even ordinary teens on Instagram agonizing over how best to project a lifestyle they don't actually have. People get different search results for the same query. Everything requires trust; at the same time nothing deserves it.

This is the one interesting thing about blockchains: they might be that one tiny gear that lets us create systems you don't have to trust. You've learned the only thing about blockchains that matters: they're boring, inefficient, and wasteful, but, if well designed, they're practically impossible to tamper with. And in a world full of shifty bullshit, being able to prove something is true is a radical development. Maybe it's the value of your bank account, maybe it's the provenance of your pair of Nikes, or maybe it's your for-real-this-time permanent record in the principal's office, but records are going to transform into chains we can't easily break, even if they're open for anyone in the world to look at.

The hype is a world where everything can be tracked and verified. The question is whether it's going to be voluntary.

BW: That got dark fast. Are you optimistic about how blockchains are going to be used once we get out of the experimental phase?

ES: What do you think? ●

A COMPENDIUM OF LAW ENFORCEMENT SURVEILLANCE TOOLS

By Edward F. Loomis

Although license plate readers did not have their origins as DHS devices, their concept and early prototype design grew from research within Britain's Police Scientific Development Branch. License plate readers capture images of vehicle tags and either retain the images locally for later forwarding and analysis or feed live images to optical character recognition software on a remote server for immediate processing. The server digitally searches for a match against a database of plates that are of police interest. The camera may be mounted on a fixed structure such as a light pole, an overpass, or a toll collection facility, in a parking lot, or on a mobile platform such as a police vehicle.

All vehicle tags that come within view of a license plate reader are captured, along with the location, date, and time of the observance and a photograph of the vehicle and possibly its occupants. Once uploaded to the analyzing server and stored, the information may be shared with other law enforcement departments.

By correlating the plate data with vehicle registration information, police officers are able to identify the vehicle's owner. When analyzed, the data can reveal personal information about innocent citizens, their travel habits and patterns, and their activities—such as attendance at a gun show or a political rally—that would be considered protected under the First Amendment.

Private companies employ license plate readers to collect data that they can sell. On May 20, 2015, Vigilant Solutions, a subsidiary of VaaS International Holdings, Inc., announced that it held over three billion license plate scans in its database and was adding new scans at a rate of one hundred million each month, offering that information to its subscribers. Vigilant's customer base is the law enforcement community, for which it tailors intelligence solutions to enhance policing efforts. Collected images may be stored indefinitely, as there is no legal requirement that license plate scans be deleted after a set period of time.

A survey of more than seventy police departments, conducted in 2011 by the Police Executive Research Forum, discovered that 71 percent of the responding law enforcement agencies were using license plate readers, that the number of departments planning to use them over the next five years would increase to 85 percent, and that on average 25 percent of department squad cars would be equipped with readers by 2016. While license plate readers have contributed to the recovery of stolen vehicles and abducted children, and helped in solving other crimes, a 2013 ACLU report compiled from the analysis of twenty-six thousand pages of data from 293 police departments in thirty-eight states warned that the readers enabled the illegal targeting of religious and ethnic minorities in the United States, and a 2012 AP article cited that the NYPD had used license plate readers in the vicinity of mosques in 2006 to profile Muslim worshipers. ●

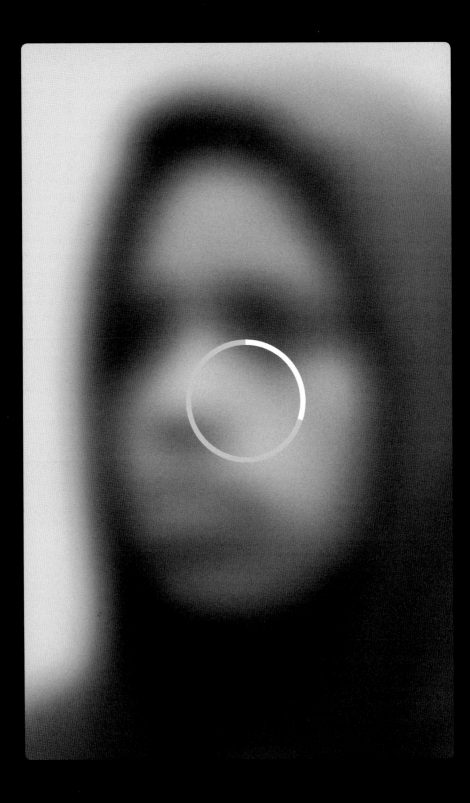

Watching the Black Body

Malkia Cyril

n December 2017, FBI agents forced Rakem Balogun and his fifteen-year-old son out of their Dallas home. They arrested Balogun on charges of illegal firearms possession and seized a book called *Negroes with Guns*. After being denied bail and spending five months in prison, Balogun was released with all charges dropped.

To his shock, Balogun later discovered that the FBI had been monitoring him for years. He also discovered that he had been arrested that day for one specific reason: he had posted a Facebook update that criticized police.

Balogun is considered by some to be the first individual prosecuted under a secretive government program that tracks so-called "Black Identity Extremists" (BIEs).

A Black Extremist is what the FBI called my mother, fifty years ago.

HISTORY REPEATS ITSELF

There were definitely extreme things about my mother. The pain of living with sickle cell anemia was extreme. The number of books she thought she could fit into a room, piling them high in the living room of our brownstone home: that was extreme.

I remember sitting on her shoulders during my first protest, in the early 1980s, against the deportation of Haitian people arriving by boat. Sitting up there, on top of my very small world, listening to extreme story after extreme story of Black bodies washed out to sea for attempting only to seek a better life, I began to understand clearly: being Black in America, and anywhere in the world, was an extreme experience.

But was it extreme to want safety, freedom, and justice for herself, her family, her people?

Despite the pain and complications of sickle cell anemia, and until the disease took her life in 2005, my mother worked every day as an educator, doing her part to achieve human rights for all oppressed people and specifically for Black people in the United States. Whether at a protest, on the floor of Liberation Bookstore in Harlem, in the darkened home of Japanese activist

Yuri Kochiyama, or at a polling site on election day, my mother always took time to tell me and my sister stories about her days stuffing envelopes for the Student Nonviolent Coordinating Committee, then as the citywide coordinator for the Free Breakfast for Children Program, operating in churches throughout New York City at the time. According to my mom, finding her voice in the Black liberation movement was powerful.

Yet, because of her voice, up until the moment of her death, my mother's Black body was also under constant surveillance by the FBI and other government agencies.

We felt the FBI's surveillance of my mother directly in the late 1970s. In order to harass her, the FBI sent my mother's file both to Health Services, where she worked as the assistant director for mental health programs in New York jails, and to the corrections officers at the jails where she worked. To their credit, Health Services rebuffed the FBI's intervention. The Office of Corrections, however, banned my mother from the jails, forcing her to supervise her programs from offsite. I remember when, years later, my mother gained access to her FBI file via a Freedom of Information Act request. It was thick, with reams of redacted pages that spoke of police and FBI targeting as far back as the mid-1960s.

Two weeks before my mother died, FBI agents visited our home, demanding that she come in for questioning about a case from the 1970s. My mother could barely walk, was suffering from some dementia, and was dying. She refused.

My mother was the target of surveillance because of her commitment to social justice and human rights. Because of this, I grew up with government surveillance as the water in which I swam, the air that I breathed.

I came to learn that the FBI has a long history of monitoring civil rights and Black liberation leaders like Ella Baker, Fannie Lou Hamer, and Martin Luther King, Jr. They monitored Black Muslim leaders like Malcolm X. They monitored Black immigrant leaders like Marcus Garvey.

I came to learn about the FBI's Counterintelligence Program, or COINTEL-PRO, the covert government program started in 1956 to monitor and disrupt the activities of the Communist Party in the United States. Its activities were often illegal, and expanded in the 1960s to target Black activists in the civil rights and Black Power movements, calling these activists–you guessed it–Black Extremists.

In 1975, a Senate Committee, popularly known as the Church Committee, was formed to investigate the FBI's intelligence programs, a response to pressure from a group that released papers exposing the existence of COINTELPRO. In a 2014 piece for the *Nation,* Betty Medsger outlines the Committee's conclusion not only that African Americans were being watched by the government more than any other group was, but that the FBI didn't require any evidence of "extremism" in order to justify the surveillance. For our communities, it didn't matter if you had never uttered a subversive word, let alone taken part in any violence. As Medsger writes, "being Black was enough." This warrantless spying on Black activists resulted in dozens of Black deaths by police shooting, and other Black lives swallowed whole for decades by the wrongful incarceration of political prisoners. Men like Kamau Sadiki and women like Safiya Bukhari, whom I grew up calling uncle and aunt, were among them.

Ultimately, the Church Committee's final report concluded that COINTELPRO was a dangerous program. As Robyn C. Spencer explains in *Black Perspectives*, the report states that the FBI used tactics which increased the "risk of death" while often disregarding "the personal rights and dignity of its victims." The Committee determined that the FBI used "vaguely defined 'pure intelligence' and 'preventive intelligence'" justifications for its surveillance of citizens who hadn't committed any crimes–for reasons which had little or nothing to do with the enforcement of law.

Given this history, my mother's history, my history, I was not surprised when *Foreign Policy* broke the story that an August 2017 FBI intelligence assessment had identified a new designation: the "Black Identity Extremist."

I was not surprised, but I was scared.

SO, WHAT IS A "BLACK IDENTITY EXTREMIST"?

"Black Identity Extremist" is a constructed category, invented by the FBI and documented in an August 2017 assessment entitled "Black Identity Extremists Likely Motivated to Target Law Enforcement Officers."

The FBI fabricated the BIE designation to create suspicion of Black people who respond publicly to police extrajudicial killings, but it doesn't stop there. The document also focuses heavily on the convergence of what it calls "Moorish [Muslim] sovereign citizen ideology" and Black radicalization as reasons for heightened law enforcement targeting. As support, the assessment specifically cites the completely unrelated cases of Micah Johnson, a man who shot and killed multiple Dallas police officers during a protest in 2016; Zale H. Thompson, who attacked police in Queens, N.Y., with a hatchet in 2014; Gavin Eugene Long, who murdered multiple police officers in Baton Rouge, La.; and a few other unnamed subjects. In each of these cited incidents the perpetrators acted alone and without any connection to each other beyond the fact that they were all Black men. Not only are these cases unrelated to each other, but they are all unrelated to the larger organized movement for Black lives in general and the Black Lives Matter Global Network in particular. The FBI's goal is clear: to fictitiously link democratically protected activities aimed at ending police violence and misconduct with what it calls "premeditated, retaliatory, lethal violence" against police officers. This is not only unethical and unaccountable; it places Black lives in real danger.

Even the FBI's own definition in the assessment is vague and likely unconstitutional: "The FBI defines black identity extremists as individuals who seek, wholly or in part, through unlawful acts of force or violence, in response to perceived racism and injustice in American society, [to establish] a separate black homeland or autonomous black social institutions, communities, or governing organizations within the United States." This definition–encompassing any act of force conducted, even partially, in response to injustice in society–has no limit. It gives the FBI and prosecutors broad discretion to treat any violence by people who happen to be Black as part of a terrorist conspiracy. It is also absolutely baseless.

The fact is, as the *Washington Post* reported in 2015, police officers are no more likely to be killed by Black offenders than by white offenders. More than half of all officers killed in the line of duty die as a result of accidents in the commission of their job rather than attacks of any kind. The total number of officers killed in the ambush-style attacks that are central to the BIE narrative remains quite small, and overall recent officer deaths remain below the national average compared with the last decade.

The bottom line is: "Black Identity Extremists" do not exist. The FBI's assessment is rooted in a history of anti-Black racism within and beyond the FBI, with the ugly addition of Islamophobia. What's worse is that the designation, by linking constitutionally protected political protest with violence by a few people who happen to be Black, serves to discourage vital dissent. Given the FBI's sordid history, this assessment could also be used to rationalize the harassment of Black protesters and an even more militant police response against them.

Despite the grave concerns of advocates, the FBI assessment and designation are already being used to justify both the erosion of racial justice-based consent decrees and the introduction of more than thirty-two Blue Lives Matter bills across fourteen states in 2017. The FBI's assessment also feeds this unfounded narrative into the training of local law enforcement. A 2018 course offered by the Virginia Department of Criminal Justice Services, for instance, includes "Black Identity Extremists" in its overview of "domestic terror groups and criminally subversive subcultures which are encountered by law enforcement professionals on a daily basis."

THE HIGH-TECH POLICING OF BLACK DISSENT

The BIE program doesn't just remind me of COINTELPRO; it represents its reemergence, this time in full view. Today, though, aided by the tech industry, this modern COINTELPRO has been digitized and upgraded for the twenty-first century.

Just as Black Lives Matter and a broader movement for Black lives organize to confront persistently brutal and unaccountable policing, programs like

BIE are legalizing the extrajudicial surveillance of Black communities. Police access to social-media data is helping to fuel this surveillance. Big tech and data companies aren't just standing by and watching the show; they are selling tickets. And through targeted advertising and the direct sale of surveillance technologies, these companies are making a killing.

Too many people still believe that civil and human rights violations of these proportions can't happen in America. They either don't know that they've been happening for centuries or wrongly believe that those days are long over. But right now, American cities with large Black populations, like Baltimore, are becoming labs for police technologies such as drones, cell phone simulators, and license plate readers. These tools, often acquired from FBI grant programs, are being used to target Black activists.

This isn't new. Tech companies and digital platforms have historically played a unique role in undermining the democratic rights of Black communities. In the twentieth century the FBI colluded with Ma Bell to co-opt telephone lines and tap the conversations of civil rights leaders, among others.

Given this history, today's high-tech surveillance of Black bodies doesn't feel new or dystopian to me. Quite the opposite. As author Simone Browne articulates beautifully in her book *Dark Matters*, agencies built to monitor Black communities and harbor white nationalists will use any available technology to carry out the mandate of white supremacy. These twenty-first-century practices are simply an extension of history and a manifestation of current relations of power. For Black bodies in America, persistent and pervasive surveillance is simply a daily fact of life.

In fact, the monitoring of Black bodies is much older than either the current high-tech version or the original COINTELPRO. Browne notes that in eighteenth-century New York, "lantern laws" required that enslaved Black people be illuminated when walking at night unaccompanied by a white person. These laws, along with a system of passes that allowed Black people to come and go, Jim Crow laws that segregated Black bodies, and the lynching that repressed Black dissidence with murderous extrajudicial force, are all forms of monitoring that, as Claudia Garcia-Rojas observed in a 2016 interview with Browne for Truthout, have "made it possible for white people to identify, observe, and control the Black body in space, but also to create and maintain racial boundaries." These are the ongoing conditions that gave birth to the FBI's BIE designation.

It has always been dangerous to be Black in America. The compliance of tech companies and—under the leadership of Attorney General Jeff Sessions and President Trump—the BIE designation escalate that danger exponentially.

For example, while many have long fought for racial diversity in Amazon's United States workforce, few were prepared for the bombshell that Amazon was selling its facial recognition tool, Rekognition, to local police departments, enabling them to identify Black activists. The problem is made worse by the fact that facial recognition tools have been shown to discriminate against Black faces.

When the Center for Media Justice (CMJ), the organization I direct, and dozens of other civil rights groups demanded that Amazon stop selling this surveillance technology to the government, the company defended its practice by saying, "Our quality of life would be much worse today if we outlawed new technology because some people could choose to abuse the technology." Such appeals assume a baseline of equity in this country that has never existed, ignoring the very real anti-Black biases built into facial recognition software. Amazon's response also rejects any responsibility for the well-known abuses against Black communities. But these happen daily at the hands of the same police forces who are buying Rekognition. Put simply, whether they acknowledge it or not, Jeff Bezos and his company are choosing profits over Black lives.

The proliferation of ineffective, unaccountable, and discriminatory technologies in the hands of brutal law enforcement agencies with a mandate to criminalize legally protected dissents using the FBI's BIE designation isn't simply dangerous to Black lives—it's deadly.

In 2016, the ACLU of Northern California published a report outlining how Facebook, Instagram, and Twitter provided users' data to Geofeedia, a social-media surveillance product used by government officials, private security firms, marketing agencies, and, yes, the police to monitor the activities and discussions of activists of color.

These examples show that our twenty-first-century digital environment offers Black communities a constant pendulum swing between promise and peril. On one hand, twenty-first-century technology is opening pathways to circumvent the traditional gatekeepers of power via a free and open internet—allowing marginalized communities of color to unite and build widespread movements for change. The growth of the movement for Black lives is just one example. On the other hand, high-tech profiling, policing, and punishment are supersizing racial discrimination and placing Black lives and dissent at even graver risk. Too often, the latter is disguised as the former.

DEFENDING OUR MOVEMENTS BY DEMANDING TECH COMPANY NONCOMPLIANCE

One way to fight back is clear: organize to demand the noncompliance of tech companies with police mass surveillance. And—despite Amazon's initial response to criticism of its facial recognition technologies—public pressure on these public-facing technology companies to stop feeding police surveillance has succeeded before.

To fight back against Geofeedia surveillance, CMJ partnered with the ACLU of Northern California and Color of Change to pressure Facebook, Instagram, and Twitter to stop allowing their platforms and data to be used for the purposes of government surveillance. We succeeded. All three social media platforms have since stopped allowing Geofeedia to mine user data.

Both from within—through demands from their own workforce—and from without—through pressure from their users, the public, and groups like CMJ and the ACLU—we can create an important choice for public-facing companies like Amazon, Twitter, IBM, Microsoft, and Facebook. We can push them to increase their role in empowering Black activists and to stop their participation in the targeting of those same people.

The path forward won't be easy. As revealed by the Cambridge Analytica scandal, in which more than eighty million Facebook users had their information sold to a political data firm hired by Donald Trump's election campaign, the high-tech practices used by law enforcement to target Black activists are already deeply embedded in a largely white and male tech ecosystem. It's no coincidence that Russian actors also used Facebook to influence the 2016 American elections, and did so by using anti-Black, anti-Muslim, and anti-immigrant dog-whistle rhetoric. They know that the prejudices of the general public are easy to inflame. Some in tech will continue to contest for broader law enforcement access to social-media data, but we must isolate them.

The point is, demanding the non-cooperation of tech and data companies is one incredibly powerful tool to resist the growing infrastructure threatening Black dissent.

In a digital age, data is power. Data companies like Facebook are disguised as social media, but their profitability comes from the data they procure and share. A BIE program, like the surveillance of my mother before it, needs data to function. The role of tech and data companies in this contest for power could not be more critical.

SURVEILLANCE FOR WHOM?
THE MYTH OF COUNTERING VIOLENT EXTREMISM

In attempting to justify its surveillance, the government often points to national security. But if the FBI, Attorney General Sessions, and the Department of Justice truly cared for the safety of all people in this country, they would use their surveillance systems to target white nationalists. For years, the growing threat of white-supremacist violence has been clear and obvious. A 2017 Joint Intelligence Bulletin warned that white-supremacist groups "were responsible for 49 homicides in 26 attacks from 2000 to 2016... more than any other domestic extremist movement" and that they "likely will continue to pose a threat of lethal violence over the next year." Yet little has been done to address this larger threat.

A heavily resourced structure already exists that could theoretically address such white-supremacist violence: Countering Violent Extremism (CVE). These programs tap schools and religious and civic institutions, calling for local community and religious leaders to work with law enforcement and other government agencies to identify and report "radicalized extremists" based on a set of generalized criteria. According to a Brennan Center report, the criteria include "expressions of hopelessness, sense of being unjustly treated, general health, and economic status." The report points out that everyone from school officials to religious leaders is tasked with identifying people based on these measures.

Yet despite being couched in neutral terms, CVE has focused almost exclusively on American Muslim communities to date. Recently, the Trump administration dropped all pretense and proposed changing the program's name from Countering Violent Extremism to Countering Islamic Extremism. As reported by *Reuters* in February 2017, this renamed program would "no longer target groups such as white supremacists."

The disproportionate focus on monitoring Muslim communities through CVE has also helped justify the disproportionate focus on so-called Black extremism. About 32 percent of U.S.-born Muslims are Black, according to the Pew Research Center. In this way, the current repression of Black dissent by the FBI is connected, in part, to the repression of Islamic dissent. As noted above, the BIE designation ties directly to Islam. And, of course, CVE programs were modeled on COINTELPRO, and the BIE designation is modeled on the successful targeting of Muslim communities in direct violation of their civil and human rights. And tech is here, too. CVE works in combination with the reflexive use of discriminatory predictive analytics and GPS monitoring within our criminal justice system. Add to this the growth of the Department of Homeland Security's Extreme Vetting Initiative, which uses social media and facial recognition to militarize the border and unlawfully detain generations of immigrants. Together, these programs create a political environment in which Black activists can be targeted, considered domestic terrorists, and stripped of basic democratic rights.

WE SAY BLACK LIVES MATTER

The FBI's BIE designation was never rooted in a concern for officer safety or national security. It wasn't rooted in any evidence that "Black Identity Extremism" even exists. None of those justifications hold water. Instead, it is rooted in a historic desire to repress Black dissidence and prevent Black social movements from gaining momentum.

And yet the movement for Black lives has, in fact, gained momentum.

I became a member of the Black Lives Matter Global Network after the brutal killing of Trayvon Martin and subsequent acquittal of his killer, George Zimmerman. It was extraordinary to witness the speed and impact with which the movement grew online and in the streets. Spurred on by the bravery of Black communities in Ferguson, Mo., I was proud to be part of that growth: marching in the street, confronting the seemingly endless pattern of Black death by cop. It was an extraordinary feeling to stand with Black people across the country as we shut down police stations in protest of their violence, halted traffic to say the names of murdered Black women, and ultimately forced Democratic candidates to address the concerns of Black voters.

The FBI's BIE designation is a blatant attempt to undermine this momentum. It seeks to criminalize and chill Black dissent and prevent alliances between Black, Muslim, immigrant, and other communities. While Black activists may be the targets of the BIE designation, we aren't the only ones impacted by this gaslighting approach. Resistance organizers working to oppose the detention, deportation, and separation of immigrant families; those fighting back against fascism and white supremacy; Muslim communities; and others are being surveilled and threatened alongside us.

In 2018, we have a Supreme Court that has upheld an unconstitutional Muslim ban alongside White House efforts to deny undocumented families due process; we have an Attorney General and a Department of Justice that endorse social-media spying and high-tech surveillance of people for simply saying and ensuring that Black lives matter. It's no coincidence that as Black activists are being targeted, the House of Representatives has quietly passed a national "Blue Lives Matter" bill, which will soon move to the Senate, protecting already heavily defended police. This even as the victims of police violence find little justice, if any, through the courts due to the thicket of already existing protections and immunities enjoyed by the police.

The movement for Black lives is a movement against borders and for belonging. It demands that tech companies divest from the surveillance and policing of Black communities, and instead invest in our lives and our sustainability.

If my mother were alive, she would remind me that a government that has enslaved Africans and sold their children will just as quickly criminalize immigrant parents and hold their children hostage, and call Muslim, Arab, and South Asian children terrorists to bomb them out of being.

She would remind me that undermining the civil and human rights of Black communities is history's extreme arc, an arc that can still be bent in the direction of justice by the same bodies being monitored now. The only remedy is the new and growing movement that is us, and we demand not to be watched but to be seen. ●

#3B5998

35ml

The
Digital Blues

Jennifer
Kabat

O n a beach as a teenager, the artist Yves Klein and two friends decided, like Greek gods, to divide up the realms. One got the sea, another the earth, and Klein the sky. Here by the Mediterranean's shores, he complained about the gulls stealing his vision, tearing through the azure overhead. "I began to feel hatred," he wrote later, "for the birds which flew back and forth across my blue sky, cloudless sky, because they tried to bore holes in my greatest and most beautiful work." Soon the sky became his domain. He went on to trademark a shade of blue. International Klein Blue was modeled on lapis lazuli. In 1957 he declared that we'd entered the Époque Bleue, the blue age. I think of him now, claiming the sky, hating the birds, calling that blue above his best work, as I consider different skies, different blues, different realms. These too are divvied up and doled out. They appear on the internet and our computers and phones, where blue is the most common color, as if that blue, the époque bleue, has only just dawned.

I started pondering this blue age as an accident. I stared at my screen, a proverbial window, facing an actual window as I stood at my desk. I was writing, hence easily distracted. I noticed the icons for different apps lined up in the dock, as Apple calls it. Scrolling left to right, there was the Finder's square smiley face, a cubist rendering in two blues, the face split like Picasso's *Demoiselles*, shrunk to a half-inch square. The App Store, iChat, Mail, and Safari veered close to Klein's luminescent lapis. Microsoft Word was turquoise with shadows in the folds of the *W*, as if to trick me into believing it had dimensionality, and the few apps that were white or gray took on a blue cast from the LED screen. On the document I wrote in, the formatting was blue, the margins marked in blue. An hour later I started an email, and the recipients' names came in two different blues: a pale one indicating the To line, while, as I typed the email address, Mail helpfully offered a list of names in navy to jog my memory. And online, links were blue. I returned to my writing but couldn't escape the blues. I highlighted text to delete it and start my sentence over, and this, too, was the same aching blue of a winter's sky in the Catskills, where I live.

I tried to work, told myself not to think of these hues, but the blues did something to me. I couldn't shake them. It was the spring of my mother's death, and distraction came easily. Research was the only thing that soothed me, as if getting lost in ideas was my salvation. My sister was sick and another friend was, too, both with cancers that had spread. It felt like my world was engulfed in waves of grief. Perhaps my blue was one of sadness, though I hate it when emotions are attached to the color, as if that might explain its grip. Blue held me in its sway through all the seasons from spring to summer to fall and the following spring again. Someone told me this ether of screens that could suck up attention was called the blue nowhere. I googled the phrase, and the string of search results turned up in blue. It's the name of a thriller by Jeffery Deaver. I didn't want to share his blues. Still, the shade I saw was foreboding, the gathering dark as the sky settled to night. In this vision, the clouds were black, and the mountains, too, at the horizon, but the sky was a deep, deep blue. Illuminated from behind, it was vivid and inky while the rest of the landscape had sunk into dusk.

Another artist, Derek Jarman, equated blue with death and loss–his own death and loss. In 1993 he made a film titled simply *Blue*. The movie is one uninterrupted royal blue shade for eighty minutes. He'd partially lost his vision to complications from AIDS and the vision he had left was tinged by the movie's color: blue. In it Tibetan bells chime and voices speak; there's ambient sound from a world he was losing: streets, cafés, doctors' offices; the whir and click of an eye test as the machine measures his retina. The experience is intensely intimate as the screen is reduced to a single field of vision. "Blue transcends," Jarman's rich voice intones, "the solemn geography of human limits." *Blue* had been his final film, and I watched it on YouTube. It'd had more than 104,000 views, with 463 thumbs up and 11 down. I clicked a thumbs up, and it, too, was blue.

I was convinced something more lurked in the shades, something perhaps prophetic. Onscreen, they beckoned and also seemed to hope I might miss them entirely, which seemed to be the point of blue–to appear and disappear, as if it were the color of nothing and everything. The color might just be a bit of digital detritus or marketing. What was the difference, after all, between these virtual blues and the ones in the "real" world, where the color dominated corporate logos and those of Major League Baseball teams?

This blue, this sky, this screen-as-window, has a nearly universal reach thanks to computers. The colors, however, come largely from one small sliver of the world's population, in one small sliver of the world on the West Coast of the United States. Those facts of place and people started to seem prophetic, too. Friendship is blue, and our language for images is watery. They come in streams, torrents, and floods. We have an image stream, video streams, and, on the iPhone, a photo stream. Even that dock, where the apps line up so they are easier to find, suggests that the programs are moored boats waiting to take out on the water. Meanwhile our data is in the sky, in the "cloud." We have clouds and currents, streams, skies, and windows. And blue.

The internet dates to the late '60s and the Cold War. Developed by an agency within the United States Department of Defense called the Defense Advanced Research Projects Agency (DARPA), the idea had been to link computers in case of war, but also to connect universities and knowledge and ideas. Knowledge, war, freedom, and information had been at the internet's heart. Technology is never neutral. It always bears out the biases of its moment. This was why I wanted to examine blue, to slow down enough over something that might seem insignificant. The color would have been easy to ignore, yet it now literally underlines our maps and paths of the world with highlighting and links as the color extends our ideas, networks, and commerce. The internet is our new civic realm, and things there are invisible because we made them so. The internet has shaped our interactions, and here was a color that had become intrinsic to them.

FACEBOOK—HEX: #3B5998; RGB: 59, 89, 152

Dusty, dark, the skin of a ripe blueberry or my need for distrac-tion. Its lowercase f sits off-center, the top curling, beckoning like a finger saying "come here." I nearly left it on my birthday when the site exhorted my friends to "help Jennifer Kabat celebrate their birthday," but I didn't. Or couldn't. Or wouldn't.

Facebook's thumb, sticking up in a shade of dusty indigo, nagged at me, as did the news about LED screens. They glowed blue dangerously. Meant to mimic daylight, LEDs steal sleep and interrupt circadian rhythms, stopping the production of melatonin, creating problems with memory and insomnia. Cancer is blamed on this blue light, and death and disease were inflecting my worries, though I didn't blame blue for these hazards. Instead it was the values smuggled in with the color that disturbed me. British pollsters YouGov undertook a survey in 2015 and claimed blue is the world's most popular color, picked nearly twice as often as the nearest competitor by respondents from Britain to Germany, Australia, China, Indonesia, Malaysia, Thailand, and beyond. Of course, the survey was online. The internet age, our blue era, is so short that it's possible to see the history unfold before our eyes. Moments that seem far off happened only five years ago. Facebook is just a decade and a half old, so maybe it was also possible to untangle blue's meaning before it congealed into accepted norms.

Yves Klein had made blue his quest, seeking something spiritual in the color. Achieving it had been a struggle since the Renaissance. The shade had required binders that diluted the intensity, and the color, ultramarine, was expensive. Lapis lazuli, from which it's made, came from a remote region of Afghanistan. The pigment's price had been as stable as an ounce of gold, and until a synthetic blue was created in the early nineteenth century only the richest patrons could afford it. Maybe what we were witnessing was the same migration that happened to other colors and pigments with trade across borders and markets. Ultramarine had moved west and north with the Crusades to Europe, and cochineal, a rich crimson made from crushed beetles in the Americas, became a valuable commodity in the sixteenth and seventeenth centuries. It had changed the tastes of Europe's wealthy and eventually returned to the Americas in the red coats of the British army.

I remembered hearing a story on NPR about blue and global politics. Online, I found the segment and listened again as the *Morning Edition* host asked what the color blue meant to me. "Is it sad," she said, "or soothing, trustworthy or cold?" The burr of her voice reverberated, and in her list were hints of how people perceived the color. Her sentence arced up at the end of her introduction to tell me that in Ethiopia blue was now the color of resistance. A reporter, a man, in Addis Ababa, explained it was the name for the opposition Blue Party, chosen because it was the color of freedom and Twitter and Facebook. Social media was still uncensored in the country. His comments were all the proof I needed that blue bore symbolic values which we were exporting.

A friend, a poet, asked what my issue was with blue. She pointed out that I was wearing a blue sweater, blue jeans, and a blue down jacket. I didn't have a problem with the color per se, I said, but with what I thought it contained. British painter Chris Ofili has said, "Blue had a strength other than color strength." I realized I was being driven by what he called "the blue devils."

I listened to the NPR story over and over. The eager journalist was always bright and cheery as he reported on a woman's blue pedicure and scarf and got around to blue being the color of social media. It took hearing the segment a half dozen times to realize the American reporter had been the only one suggesting it was the opposition party's color because of social media. It was just him and me conflating color and cause. No one he interviewed did. The Blue Party spokesperson gave the reason as the Blue Nile and the Red Sea, which appeared turquoise.

Each time the NPR announcer asked if blue made me sad, her voice settled in my chest, and I thought of my father. He'd written to Adlai Stevenson in 1952, the day after Stevenson lost his first presidential bid. I found a blue carbon copy of the letter in my mother's files. My dad was twenty-six and wrote that he was sad about Stevenson's loss. He was scared of war and of returning to active duty. At the time he managed a tiny electric co-op in upstate New York, and he worried, too, that the state would privatize its big power projects in Niagara Falls and the Saint Lawrence River. He believed these resources belonged to the public. He believed in the public good, not privatizing infrastructure. This was the thing with blue–I was sure it was bound up with privatizing something that should be a public resource.

TWITTER—HEX: #55ACEE; RGB: 85, 172, 238
Bold azure, the endless sky on days of weight and ache.

I called Michael Bierut, the designer who created Hillary Clinton's 2016 presidential campaign logo, to ask about blue, and he sent me to Jessica Helfand for answers. Together, they cofounded Design Observer, a site dedicated to thinking broadly about design and culture. She teaches a class at Yale on the color blue. Yale's color is blue, and she told me a story of the day Twitter launched. She'd been at a design conference where a Twitter spokesperson introduced this little bird, the bluebird of happiness. It was a jaunty aqua, and she thought the whole thing ridiculous. "Typing a hundred and forty-character messages?" She gave me a self-deprecating laugh.

Jessica also offered up a story of Paul Rand, as if by example. The legendary designer had created logos for IBM and ABC, and tried to turn American Express a teal nearly the color of Tiffany's. "It wasn't exactly Tiffany blue," she related, "but this was before the digital age, so people couldn't track down your hexadecimal code[1] and say, 'You stole my swatch.' He was alluding to that very bright shade of robin's-egg blue, and felt that he was psychologically importing the value proposition. It was the color of money. It said wealth, it said exclusive. He was borrowing the cumulative cultural legacy of that memory." She also hinted that he'd stepped very close to the line between borrowing and maybe something more. Perhaps this was blue, borrowing a legacy and bordering on theft? Maybe all these logos were just built on the color's previous uses and the myriad values we think blue has, from sadness to freedom and peace and money.

Amex now comes in two blues: one darker, a serious navy shade; the other lighter and closer to Tiffany's. Skype and Twitter use a similar hue. Skype's is in the shape of a cloud, Twitter's the bird, and Jessica said, "Blue is the path of least resistance. These blue brands," she told me, "aren't worthy of your efforts. These visual decisions are just the result of external consensus-building."

I could feel her telling me to give up my quest. She said I was a dog with a bone. A blue bone, I joked, or blue blood. The color might be easy to shrug off, but that's why I wasn't going to.

---

1. The six-digit alphanumeric code that defines a specific RGB color used on the internet.

A friend at Apple, who couldn't legally discuss with me the blues like Mail or Safari or even Finder that his company had created, sent me to another creative director, Aki Shelton, who used to work at the company. Connected by Skype's cyan cloud, she told me about working for a design agency in Japan. One of her clients was a bank in Taiwan. "In Japan," she said, "blue never had negative connotations, but for the bank I was making a logo mark, and it was blue, a sky blue they called 'dead man's face blue' as if it were the color of death." She explained that in China and Taiwan people take the fact that red and gold represent good luck very seriously.

She had been thinking, however, of a different meaning for the color. "In the U.S. and UK, it's trustworthy and friendly, safe and modern—all these make it popular. Health-care companies and health insurance providers are blue." [2]

She leaned toward her screen and toward me and said another thing about blue: "It recedes." This was why paintings used blue to convey distance. With atmospheric perspective, contrast decreased and everything blended into the background color—blue, most often.

"When looking at Facebook," she said, "that blue basically disappears. Facebook is about users' content and photos. They should stand out so the color should step back, and social media uses blue for that reason."

The color disappears. Blue is the color you don't see, the color of neutrality but also safety and trust. Perhaps because blue is so ubiquitous, it can represent all of them or nothing, just neutral space. Perhaps blue can represent those values because it's so common it's invisible. At the same time, I think its familiarity renders it trustworthy and reassuring. We see the color so often, it doesn't jar us. Blue is comfortable.

Just before I got off my Skype call with Aki she mentioned that she'd recently created a blue identity—it was a dusty indigo and she used it for the Public Internet Registry. It's the nonprofit responsible for all .org and .ngo domain names, most often associated with nonprofits. Their shade? "Blue," she said, "for trust."

---

2. Aki also explained that her team knows how common blue is online and tries to point it out to clients, who don't necessarily listen to their designers' advice.

SKYPE—HEX: #00AFF0; RGB: 0, 175, 240

A cyan daydream, hidden in a cloud and promising conversations with anyone anywhere, as if talk were cheap.

Anytime anyone told me blue was the color of *something*–of, say, trust or peace or calm–I got suspicious. I'd wanted to tell the NPR host that, yes, I was sad, but it had nothing to do with color. Color as feelings seemed too facile, like the results from an online psych quiz:

> I am a:
> ○ Male ○ Female
>
> My future seems hopeless:
> ○ Not at all
> ○ Just a little
> ○ Somewhat
> ○ Moderately
> ○ Quite a lot
> ○ Very much

A creative director from ABC News who had also been in charge of its digital platform told me blue was "authority." It was why news outlets used it. She also mentioned that red couldn't be used well on screens until recently. You couldn't reproduce it without "bleeding." It had been too hard to control until the new retinal displays were developed. "Social networks have built on the legacy of blue, on the trust from news organizations," she explained.

Like Rand, they had been borrowing–or stealing–associations. So red was tricky; blue was trust and authority. And credit cards and commerce, news outlets and technology, are all using a color to represent these ideals. There's something unnerving about the way the notion of trust (and specifically what news organizations and tech companies and financial institutions consider trust) is derived from simple consensus-building that can be summed up in a color. The feedback loop reinforces blue's ubiquity, so we see it more and more, become more and more accustomed to it, and that repetition translates into trust. We don't have to work to understand blue.

Perhaps what disturbs me more is how this twinning of trust and authority connects to the ease of consumption. Blue serves as shorthand, a signal of trust. Thus the news story is easier to consume, the social-media platform more familiar. The more comfortable we feel on it, the more likely we are to put our data and our political decisions into a site. Companies seem to think that all we need to trust something is to be shown it repeatedly—no matter its actual relation to fact or security.

DROPBOX—HEX: #0061FE; RGB: 0, 97, 254

Egyptian blue, close to the hue recreated by chemist and former slave George Washington Carver around 1930. He made it from mud, from clay... And, under this shade, I store things, share and trade them with friends: images, words, their art, my writing, believing these will all last. Or float free.[3]

My husband, a graphic designer, overheard these conversations. He'd been listening to them for months and finally told me he thought the blues had a longer and deeper history, one that wasn't about feelings or abstract ideals, one that was more about how colors work on screens. The blues came, he suspected, from the web-safe color palette. It was a name I'd not heard for a decade. The palette had been the set of colors for the internet that could reliably be reproduced on both Macs and PCs. Both systems could display their own set of 256 shades. They had 216 of these shades in common, which were therefore deemed trustworthy. Of those, twenty-two were truly reliable. Stray from them and specify a different color, he explained, and it would "dither." That was the hatched pattern of visual noise combining two different shades like red and yellow to approximate the one you'd chosen. He pulled up the color palette

3. And I owe knowledge of this blue and Carver to the late Terry Adkins, whose installation and recital *Nenuphar* compared Carver and Yves Klein and their two blues.

on his screen and the hues were garish: acid lime greens and yellows, purples that veered to neon.

On a hike that summer, I was still harping on blue. The hill and trees were lush and green. Ferns brushed our legs and the blackberries were starting to ripen. We climbed to a fire tower to see the atmospheric blues scattering light and creating a sense of distance, that blue of perspective Aki had talked about, which appeared and receded. Before we reached the top, my husband said that the web-safe color palette had always looked random to him. "There was no logic to it. It didn't conform to nature or skin tones. The colors would go from really saturated to really dark, none of which was helpful as a designer. It was a cube of colors that was a mathematically derived formula, so they don't correspond to nature, and in a way that is amazing."

What he meant in calling it amazing was that it wasn't biased towards skin tones, unlike technologies like printing that, when aiming to recreate flesh colors, skewed white with an implicit racism. He hoisted his backpack. "But the palette was designed to depict every physical color within 216 mathematical divisions. So it was perfectly inhuman, but it also meant there were few attractive colors, and we often would try to find clever ways to make our own, creating our own dithered patterns of stripes and hatches to modify a shade. Or use blue."

WORDPRESS—HEX: #21759B; RGB: 33, 117, 155

Inky blue, it's the flash of a magpie's wings. A common bird, a corvid, it's as smart as an ape and uses language, too. It speaks and shares grief, tools, and emotions. The bird's name is a compound noun: mag, short for Margaret and once used as a synonym for all women, and pie, meaning pointed, like its beak and tail. It was the pointed bird that chattered like a woman. Its words were cheap, everywhere, chatter... Language that has become chatter, words made common, everywhere, every day.

The web-safe color palette was created for Netscape Navigator, the first commercial browser. It launched at the end of 1994, its logo the aqua of the North

Atlantic. The company's first press release declared that Netscape would sponsor the altruistic free flow of information–and money.

"Making Netscape freely available to Internet users is Mosaic Communications' way of contributing to the explosive growth of innovative information applications on global networks," said Marc Andreessen, the company's vice president of technology, more than two decades ago. He was twenty-three years old and had recently graduated from the University of Illinois, where he'd co-created Mosaic, the first popular web browser. It had been funded by the federal government, and when he made it there'd only been twenty-six websites worldwide. He moved to California afterwards and started Mosaic Communications to create a commercial browser, Netscape.

"We expect Netscape's ease of use," his statement continued, "to spark another major leap in Internet usage... Netscape now lays the foundation for commerce on the net."

By the time it was available that December, the browser was no longer free. Not even half a year later, Bill Gates pronounced in an internal memo, "Now I assign the internet the highest level of importance." Up to that point, Microsoft had had only six people working on browsers. After Gates's memo the company went on a hiring spree, and soon my friend Matt became part of the multitude working on the internet at Microsoft. We met around that time, a year or so after he'd started at the company.

The internet had been hailed as a utopia where we would express ourselves, but it was also a realm of competition and capital. Andreessen later wrote for the *Wall Street Journal* about how software companies working online "invade and overturn" and "disrupt" to become the dominant force. I thought of that "invasion" with Gates's intense desire to own the internet. Microsoft Windows 95 launched in August 1995,[4] and the software soon came bundled with Microsoft's own web browser, Internet Explorer. Its early versions were identified by a small blue marble of the earth over which hung a tiny magnifying glass.

---

4. Windows's logo was a window flying across a blue sky. A rainbow filled the panes, and a contrail of colors followed behind. The identity looked like NEXT's rainbow cube,

That earth was quickly replaced with a lowercase cerulean *e*. Microsoft's Internet Explorer had become the earth.

Microsoft's browser killed off Netscape. In what may have been the biggest irony, Microsoft licensed Andreessen's first browser, Mosaic, in order to jump quickly into the browser market itself, then poured millions into starting a browser war. Internet Explorer was essentially Mosaic, and it was pitted against Netscape, which never recovered. Meanwhile, Internet Explorer, with its little blue *e*, took over the world. Perhaps it was this blue, the globe, the marble of the earth, that became a brand that killed off the competition and tried to dominate what and how we see?

SAFARI—HEX: #1C9BF7; RGB: 28, 155, 247

Nautical blue. With a compass in the middle, it has promised me the world or a web or both, that I might find direction.

In the fall Matt and I met at a café in Soho. We sat outside. He was now an artist and a creative director and had recently quit his job for Ace Hotels. Our breath puffed in large clouds. On his lap was a tiny dog named Biggie with the face of Yoda. Matt had started at Microsoft soon after he'd graduated from the Rhode Island School of Design. He said back then Microsoft was hiring everyone to keep them all away from the competition. "The Tetris guy, who invented the game, they paid him not to work for anyone else, and they paid the Rolling Stones to use 'Start Me Up' in Windows 95's first ad campaign and Brian Eno $35,000 for the three notes you'd hear when you booted up Windows."

Microsoft launched MSN (the Microsoft Network) as its version of AOL, and Matt created a youth-culture zine for it called *Mint*. (I wrote for *Mint*, too.) "We had this idea," he said, Biggie shivering into his coat, "that we'd flip the Microsoft logo upside down. The brand people came in and said no. So, we asked again." He'd been in his early twenties and hadn't realized a brand identity

---

as if rainbows summed up the possibilities of technology to this point. Of course, Apple had a rainbow, too. But Microsoft's window linked it to blue.

was sacrosanct–that it was the vehicle for the company's values. "In all this we were being lectured about the brand guidelines, and what they meant and why. They said the brand's packaging had to be blue because blue was the most popular color." He shook his head. The moment stuck with him for more than two decades. "Blue was the world's most popular color," he repeated wistfully.

Another friend, who'd worked at Microsoft's ad agency on the launch of Windows 95, concurred. The blue hadn't been about beauty or function. The reasoning hadn't even been as good as blue skies, open windows, and freedom. It was reductive. Everything at Microsoft was data-driven, he explained. "They'd give you three options and ask which one you preferred. They didn't make decisions. They responded to data points."

This agency employee, my friend, didn't want his name quoted. He still works in advertising, still works for tech companies, and said, "Microsoft used quantitative research testing, delivered by companies that... link the data to preexisting normative data in order to predict the values. It's an egregious misuse of data." Essentially, Microsoft was making decisions on data that other companies collected and analyzed. My friend linked this back to Robert McNamara, the Ford executive who'd made design and manufacturing decisions based on data, and how McNamara used this data-driven decision-making process to measure success in Vietnam when he became Secretary of Defense. "By 19-whatever-it-was, McNamara's data and his much-vaunted computers in California had models saying we were six months away from winning the war and exterminating every Vietnamese human being." My friend swallowed a bitter laugh.

In World War II, McNamara's statistics were responsible for decisions to firebomb civilians in Japan, and they turned Vietnam into war by body count. In both cases, numbers were supposed to rationalize war, making it logical. Instead, they dehumanized everyone. Soldiers became nothing more than killing machines; massacres like My Lai resulted, and everyone the United States killed was recorded as Vietcong and North Vietnamese, regardless of their affiliation. The statistics were reported weekly in United States newspapers. While he was Secretary of Defense, McNamara's data was often so complex it was impossible to contradict. It built a wall around his decisions. No one could question it, and the statistics seemed to remove human fallibility.

My friend said it showed how wrong data can be. At best you get blue; you get algorithms making decisions.[5] The thing is, though, technology doesn't exist ex nihilo. It's created not in a vacuum but by humans, repeating our biases, our prejudices. The algorithms deployed by companies like Google or Facebook allow for only a seeming objectivity, and when a company trots out claims of objectivity, it cloaks them in a color, blue, to reinforce that image.

GOOGLE CHROME—HEX: #418BF3; RGB: 65, 140, 243

A blue dot encircled in a swirl of red, yellow, and green, like a
camera shutter closing on it, or an iris. A similar blue dot is at
the Firefox logo's center. The flame-red fox spins, trying to catch
its tail as it chases a globe. The globe, the earth, that marble,
has not disappeared from web browsers even after twenty years.

"Blue is the richest color for me," Mark Zuckerberg told the *New Yorker* when Facebook was going public. "I can see all of blue." And according to Zuckerberg, this was why Facebook is blue. His classmate Andrew McCollum created the logo in 2004 when Zuckerberg was launching the social network and made it blue on blue. The web-safe palette still held, and, for the identity, he paired a denim shade with Yves Klein's bold one. The network's name was The Facebook, which appeared wrapped in brackets as if a whisper or an aside. The site was just that at first, accessible only to students at Harvard and a few other elite universities. Zuckerberg had likely chosen blue because he was colorblind, something he discovered only years later in a test online.

---

5. In 2009, Google tested forty-one different blues to see which people used most. That same year, Microsoft launched its search engine, Bing, to compete with Google, believing the correct blue for links would translate into $80 million of revenue from searches—a paltry figure compared to the $200 million Google pulled in from their new shade of blue. For a company built on data-point tracking like Google, the 1 percent who click one shade of blue over another is mechanistic, not humanistic.

Like him, about 8 percent of men with Northern European ancestry have red-green colorblindness.

LINKEDIN—HEX: #007BB6; RGB: 0, 123, 182

Almost royal, not quite navy, not bright, not receding. There, it's a blue that is just there. Perhaps this is what they mean by corporate blue and why I never use LinkedIn–ignoring requests to "connect" or reports the site sends me–helpfully, hopefully–telling me I've appeared in three searches this week, or nine. What kinds of jobs are there for writers who lose a year to the color blue?

Maybe I was trying too hard. Maybe blue was just popular or signified trust; maybe people did copy others; maybe designers were simply at the behest of executives. The idea, though, that blue receded, that it was the distance, that it was meant to be the unseen element like formatting, as Aki had said, made me think there was something more. But some of the very first web pages, designed around 1989, used green for links. Blue wasn't always the default. Convinced that there had to be an explanation for the color's continued dominance, I called a digital curator at the New Museum who archived early websites, and a professor at Stanford, an expert in interface design who'd advised Larry Page as he developed Google. They both said there was no logical reason for blue. "It's not functional," said the professor, Page's advisor, "but fashion."

I still couldn't give up the idea of functionality. Or at least I wouldn't. There had to be an explanation. Google itself had spoofed all the blue online a few years ago, launching something called Gmail Blue as an April Fool's joke. It was a version of Gmail where everything had been turned blue. All this blue seemed problematic given Silicon Valley's goals. Companies there, as Andreessen wrote, quested after "disruption." They wanted technologies that altered behaviors or enabled new ones, new ways of communicating or shopping or having friends. If you wanted to disrupt, to build a revolution, why use a color that might remind us of all the old ways of working and thinking and talking? In that case, it seemed like there must be more to the perception of blue, like how red couldn't be rendered well on monitors. What if the choice of blue was about us, about how we see? Maybe blue got at the nature of vision itself.

My knowledge of vision ended at high school biology, where I'd learned that the eye sees in three frequencies, red, green, and blue, and that the brain then processes those three colors to trick us into perceiving multiple shades across the spectrum. That's been the accepted science since 1801, and it was the paradigm used for all screens and displays from TVs to cameras and phones. They were all still based on RGB some two hundred years later. At NYU, where I teach, I turned to Eric Rosenthal, a vision expert who's worked with DARPA as well as Disney. He promised he could help. I met him in his university office. It was in an old manufacturing building on lower Broadway where women had once sewed clothes in sweatshops, and I was led past warrens of desks with low light and glowing screens.

In a room with plate-glass windows, a workbench was covered by tackle boxes filled with circuits that twinkled like rubies. Eric introduced me to his collaborator, Richard Solomon. The two made a strange pair. Eric was elfin with a carefully trimmed beard and gravelly voice from the Bronx, where he grew up. Everything about Richard was askew: glasses, hair, and even his speech. He zoomed around subjects in urgent swerves of information. Both men were in blue, though–Richard in blue jeans, and Eric in a blue dress shirt, his NYU lanyard neatly tucked inside. I was excited about the DARPA/Disney connection, sure that this conjunction of war and entertainment would illuminate the conspiracy I suspected lay behind blue. DARPA had been responsible for the internet, and Disney has long used color to play on emotions. Instead, the two men overturned everything I'd ever learned about vision.

In the early '90s they'd worked on separate projects developing digital TV and cameras, and neither knew the other. Eric was Vice President of Advanced Technology Research at Walt Disney Imagineering Research and Development, Inc.; Richard had co-written a book on information infrastructure, and both were working on projects at MIT. Someone at the NSA thought the two should meet. Both have worked for what Richard called "the three-letter agencies" ever since then.

At the time the Department of Defense had set a project, a crucible of sorts, for the two of them to make better television, with even higher resolution than HDTV. But, I asked, what did the military want with TV?

They explained it was for analysis. "With analog film," Eric said, stroking his beard, "it took three days to process and land on an analyst's desk, but digital went directly there. Only, after HDTV"–DARPA had co-sponsored the research–"the resolution still wasn't good enough for analysts to read the images in enough detail, so they asked us what we'd do, and we said we'd rethink vision entirely."

Their quest had perhaps been as dogged as mine. They rejected all the accepted science around vision. It wasn't RGB. All of that had been wrong, they insisted, and in great wandering asides Richard explained why. Their process took them across Europe and the United States to Newcastle and Dusseldorf, London, Berkeley, Stanford, even the hippy enclave of Ojai, California.

Outside Eric's office, the sun was setting, and I was sure we'd never get to blue. The men were talking about how goldfish see infrared and how the skin behind our knees is light sensitive; how reptile skin is also light sensitive, and how frogs' vision is linked to their tongues, and how our eyes are aware of color even before the brain begins to process it. Finally, though, they got to blue. "The interesting thing about blue perceptually," Eric said, "is that we have the worst color resolution for blue and the best resolution for colors in reds, oranges, and yellows–mostly oranges and yellows–so we have the least ability to define shades of blue."

Blue was different, but it didn't sound like its differences should make it a default color, not if orange or red or yellow were easier to see. They told me, too, that specific cones in the eye catch the wavelengths for specific color frequencies, but that none of them work particularly well for blue. I told them I'd never find my answer, and Eric smiled slyly.

"The eye," he explained, "has a vitreous fluid. It's an ultraviolet filter to protect the eye and also a blue diffuser. It causes the blue to be diffused so the rods detect it. The rods are all one size in the wavelength of blue. We've got twenty million rods processing blue, and they can't do anything but process blue. And then we've got all these cones, and they're trying to read detail and do color definition and chromatic separation so that we can perceive these precise changes in color, but blue is really easy to process."

"We don't have that many blue cones, so it's the rods," he said, "that capture blue." This idea is radical; it contradicts the popular understanding that blue-sensitive cones process blue wavelengths of light.

Richard jumped in. He spoke quickly, gesturing with his hands about numbers of receptors, rods and cones. Eric calmly explained, "It takes more energy to figure out any other color compared to blue. It is easier to perceive blue, and that's why we think the sensation is that blue is more calming. It takes less brainpower. It's easy to process."

To believe them and all the money DARPA had spent to send them out into the world to find a better model of vision to make a better TV to find a better way to spy, blue had a reason for its popularity. It meant I had to trust that Eric and Richard were right. I had no way of proving they were, given that their analysis disagrees with most other scientists' analysis of vision. I was not a biologist. I couldn't do more than talk vaguely about the physics of light, and blue had already possessed me beyond all rationality. I thought of Isaac Newton, who, in trying to understand color, lanced his eyeball. For centuries artists had been driven mad by blue not because it was toxic, like lead in white paint had been, but because it drove them on this quest for a single pure hue. Blue, though, was easy on the eye. It was literally in the eye of the beholder. Only problem: the color didn't always exist, certainly not as we know it.

FACEBOOK MESSENGER—HEX: #0084FF; RGB: 0, 132, 255

A sky-blue speech bubble floats like a balloon, concealing a light-ning bolt inside. Words float free, as if hit by lightning, or as if at lightning speed. The case of language now unhinged, in which friend and social and liking have slipped from their original meanings and images stand in for words. But I love the small gap that opens between a picture and its associations. It makes me think of the Surrealists, of Magritte's pipe that was not a pipe and Dali's telephone that was a lobster.

If there's no word for something, can you see it? Is there a hole in perception or experience? How do we frame that thing? Or describe it? This was my problem with blue. It hadn't always been around, or at least there hadn't always been a word for it. Linguists, anthropologists, and ethnologists have been fighting over this for nearly fifty years. Language shapes how we see. It lays out the common currency of our world; words exist because we communally agree we need them and their meanings, so if we can't talk about something together, what then? Do we talk around it? *The Odyssey* famously has no mention of blue anywhere in the book. It might be an epic about sailing over the Aegean, but for Homer the sea was "wine-dark" and the sky "rosy-fingered"–not blue at all.

Former British Prime Minister William Gladstone also wrote about this in the nineteenth century. He'd been as bedeviled as I was by blue, only Gladstone needed to understand its absence. Other cultures were also missing the color in their vocabularies. The African Himba tribe has no word for it but many words for greens, while Russian has two distinct words for blue.[6] Light blue is *goluboi* and dark blue *sinij*. The modifiers *light* and *dark* make them seem related to an English speaker, but *goluboi* and *sinij* are akin to red and yellow, two entirely different colors. In 2006 a group of scientists in Boston studied how the two words shaped perception, testing native English speakers and native Russian speakers on these blues. The volunteers had to identify different blue squares on a screen. The ones who'd grown up with *goluboi*, the Russian

---

6. These words are called basic color terms, not something like *salmon* for pink or *cerulean* for sky blue, but the most reductive of terms, *red*, *green*, *blue*, *yellow*—the simplest way to identify a color.

speakers, were faster than the English speakers at identifying the differences.

In a friends' house upstate I conducted my own study. Alina is Ukrainian, Jeff is Russian; both are artists. He leaned against the desk. Their son's toys were strewn on the floor. She swiveled in her chair, and Jeff said the blues, from *goluboi* to *sinij*, were not all that different even if the words were. "The colors are such a continuum I can't really fix what is light blue. It goes from the lightest of light shades, which are almost white, all the way to black, basically, as *sinij*. So when you say light blue, *goluboi*, it's just a point on an infinite line."

What about mixing colors? Learning to paint?

"You add white or more water," Alina explained. "The blues mix just the same."

How had they learned them as children? She couldn't remember; Jeff said, "The sky is *goluboi*, with tints of gray and greens. Dark blue is the color of the ocean or the night sky."

We were speaking in English of a Russian blue. Already the language was slipping, and I wanted to track down that untranslatable place to understand it in Russian, not Russian rendered into English. There was a hole in our blues. I didn't see the ocean as dark like Jeff and Alina did. For me it was turquoise, and maybe that difference in language shifted how we saw color.

The friend, the poet who'd wondered why I was so dogged by blue, has traveled to Ethiopia. She asked a few Amharic and Oromo speakers about the color for me. One person was a student of hers in New York, and he wrote, "The only thing I can think of, besides the opposition Blue Party, is that the word for blue and the word for sky are nearly identical: *semay* (sky) and *semayawi* (blue). There aren't any other colors that have double meanings like that." No one mentioned social media.

Blue came to English from the Norse, from *blae*, a word that meant blue-black. It was connected to death and lividity, migrating from early German, as *blâw*, to Romance languages via the Barbarians, and for years the color was associated with heathens. In northern England and Scotland, they had blue and versions of the original *bla* as *blae* and *blay*. Not blue, it was more "blah," almost as I would picture it: a dull sunless day. A friend in China wrote to tell me that in Chinese culture color stemmed from the five elements, and blue wasn't one of them. "The color associated with the Water—think winter, stillness, etc.—is black, not blue," he emailed. *Bla*, black-blue, wine-dark.

APP STORE—HEX: #0287F1; RGB: 2, 135, 241

Encircled in royal blue and scrawled onto Yves Klein's sky, an A—the shape of the anarchy symbol—is formed by a brush and pencil. Here, anarchy sells games for micropayments.

The British YouGov opinion poll that found that blue was the world's favorite color included China, which also picked the shade as its top choice. In places like the United States and United Kingdom, though, the survey found that preference broke down by gender. Forty percent of men, while only a quarter of women, preferred blue. Silicon Valley is famously male-dominated. Maybe blue was about gender? I turned to a cognitive psychologist to find out if this was true. Karen Schloss, then a professor at Brown, told me the issue of gender and color was "hairy and complicated." She'd done her graduate research at Berkeley, near Silicon Valley. She asked if YouGov's poll had shown colors or just used the word *blue*. I told her it was only the word, and she sighed. "There are particular blues women might not call blue, that they'd say were their favorite color if they were shown them"—like her own current favorite, electric blue. She picked up her laptop; the keyboard was vibrant indigo. She wore a blue scarf and was sitting in her midnight-blue dining room, an eggplant color that I thought of as nearly wine-dark.

She studies visual perception and cognition and has made color preferences her life's work. She said Homer had possibly been colorblind, or that the phrase "wine-dark" could have been a convention at the time. "The Greeks," she assured me, "definitely had a way of talking about blue, if not the word itself as we know it."

She explained that what we think of a color is shaped by all we associate with it. The more positive things you experience in a certain shade, the more you like it, which is where kids' gendered preferences come in. Pink reinforces pink with girls' toys. This meant, too, that the associative values with blue are only becoming more powerful the more it's used. She also said many studies had reliably proved blue was indeed the world's favorite color. "There are two things everyone has in common, globally–clean, pure water and the sky, and also biological waste. Water and the sky are blue, and waste, feces, vomit skew to yellow, brown, and green."

This made sense. She was a scientist and didn't think we all saw the same colors in the same ways. In Berkeley, though, where she'd done her research, she found no gender difference in color choices, while in Serbia women apparently picked pinks and purples.

To be provocative, I asked why the formatting on the document I was typing as we talked was blue, and she cautioned that her response was only a guess. "We're used to seeing and discounting the sky all the time. It's always there, and yes, it's white now, unfortunately." We were talking on a bleak winter's day. "But for the most part it's invariant, and our visual system is trained to detect differences and trained to discount things that are constant. You don't want to have to process every single thing going on at once, so if you are sitting in your dining room and a squirrel runs by the window up the tree, that is something new and different. I want to devote resources to that, so I know if that is a danger or if it's just a squirrel outside my window. This is purely speculation," she repeated. "We are used to discounting the most prevalent background, which is the sky, which is shades of grays and blues."

The sky receded and appeared. We could focus on it as we needed. On a document, the blues and grays were the easiest to see and ignore, to see without being distracted. When we met, Eric and Richard had called the brain a difference engine. It responded to change because change triggers perception. For his students, Eric used the example of staring at a cloudless sky on the beach. "Look at it for ten minutes," he said, "and you start to see black blotches, because the brain needs something to change for perception."

Yves Klein hadn't wanted anything tracking across his sky and disrupting his vision. Karen Schloss said he wanted to give people all the possibilities of the pure blue sky. She talked about his paintings' associative power. "If you put in representative information"–her hands were spread wide, as if summoning the world into her dark blue dining room–"you lose the abstract associations. He wanted to create an experience for the viewer about that spiritual possibility contained in a single hue."

SIGNAL—HEX: #0090E9; RGB: 0, 144, 233

The jaunty blue of an indigo bunting, a blue bird that is not blue, just appears as such because of a trick of the light. The "jewel-like color comes instead from microscopic structures in the feathers that refract and reflect blue light, much like the airborne particles that cause the sky to look blue," explains the Cornell Ornithology Lab. Shine light behind a feather and the blue disappears like language on Signal. That's why I joined, so my speech could disappear. I worried not only about the future of my privacy but about the actions I might take to protect my notion of freedom and trust.

In San Francisco in early spring I went in search of one Yves Klein. It was tiny, just bigger than an app icon. Its owner was the artist and filmmaker Lynn Hershman Leeson. She's worked with interaction, technology, privacy, and mass communication since the 1970s, revolutionizing digital art in the process. Her *Difference Engine* used an early robotic interface, and in the mid-'90s she created telerobotic dolls with webcams for eyes that gallery visitors controlled. Online you could see what the dolls saw, a prescient example of internet surveillance.

We sat in an Italian restaurant in North Beach, not far from where she taught at the San Francisco Art Institute. She'd promised to bring the painting. "It's the size of a stamp," she said on the phone, "but it still has the special power of that Yves Klein blue."

Inside the lights were dim, though it was bright outside, and I wanted to ask where the Klein was before any pleasantries. Sheathed in black, Lynn had a ring of keys jangling around her wrist like a bangle. Sparkling water, bread, and olive oil all materialized on the table, and I wondered how she might produce the painting. We talked about the monograph on her work that'd just come out, and she reached for her bag. I expected her to take out the book, but instead she pulled out a gold rectangle–the frame. My heart leapt. No tissue or bubble wrap protected the painting. The only thing between the Yves Klein and me was a tiny sliver of glass. The painting was luminous, and I could call the blue *royal* or *French* but describing it would diminish its power. The painting did light up our table, just as Lynn had promised. It even had serrated edges like a stamp. "Its size doesn't matter," she said. "It has this internal essence. For him–" meaning Klein–"there was this mysticism, this floating sense of being erased in blue."

I'd been waiting months to see the painting, and here we were with it in a restaurant, of all places. I cradled the frame in my hands, and Klein's blue pigment saturated my space. I didn't want her to put it away, but I couldn't hold onto it forever. I'd been in the Bay Area for a month at an artists' residency, living on a former military base and thinking about the language of Silicon Valley–freedom, individualism, and creativity. I told Lynn I'd been writing about Jack Dorsey's–Twitter's founder–libertarianism, and Twitter's blue bird. It trumpets free language; meanwhile, Dorsey's other invention, Square, processes mobile payments, making commerce ever more free and ethereal.

She slid the painting back into her purse and stowed it on the floor. There was an Yves Klein at my feet and now a beet salad before me. If I bent awkwardly to the side, I could see a hint of the frame. I explained how my quest had been going wrong. I was losing hope (a blue-hued emotion if ever there was one), and blue had driven me to the edge. I told her I'd been mourning my parents, too, as I thought about how these ideas manifested, and maybe thinking about blue had helped me deal with those emotions.

Yves Klein used blue because he wanted to contain the sky, and Lynn talked about how Derek Jarman made his *Blue* as his world was disappearing. She asked if I'd seen the movie, that it would make sense given my grief. I told her that I'd recently watched it and was ashamed to admit I'd seen it online, chopped into sections with poor resolution–the image and sound compressed, the thumbs-up and flippant comment threads reducing something profound to a simplistic gesture. That, too, I said, seemed encoded with blue.

She told me I was not crazy for spending all of this time on the color and said that the smallness of the search made it worthwhile as I dug into the values behind the shade. She linked blue to chroma-key, to blue screens, the first technology for special effects. "It was," she said, "originally blue, a little darker than Twitter." This blue allowed things to appear and disappear, so someone could seem to be anywhere in the world.

"With blue"—the keys jingled at her wrist as she reached for the asparagus between us—"I always think of the reasons he chose the color—spiritualism and the atmosphere and our sky. With technology there are global issues, ideas going through the air, and ether and air represented by sky and blue and that connectivity. My hope is that the founders of digital media, who were mainly hippies working in garages in the '60s and '70s, were also thinking this way, like Yves Klein, using blue as a method of infusing a global spiritual sensibility into technology."

She'd lived in Berkeley then, too. "It was a moment of opening up to the individual, and technology has always been in the fault lines here. TV was invented here," she explained, adding that if she hadn't lived in the Bay Area, she doubted she'd have focused on technology in her own films and art. Place, she explained, shaped the ideas formed there.

Lynn was working now on a documentary about biological computers, run by programmed cells, that could be slipped into bodies to monitor bloodstreams. Her projects consider the dark ways technology identifies us, and she told a story of meeting developers in the early '80s who were working on interactive technology. They talked of how it would track our interests to sell us things. She kept nervously anticipating its release, only the development arrived years later than she expected.

When we left the restaurant, the sky was still bold and vibrant, and I thought of this blue of ether, of disappearance and tracking and surveillance. Blue had driven me crazy. Karen Schloss described the sky I saw and didn't see, and Yves Klein had trademarked the blue of his sky so no one else could claim it, yet it was here before me on a bright afternoon in a city that shaped the tech industry.

I called Rob Janoff, one of those hippies Lynn had talked about. He designed Apple's rainbow-striped logo in 1976. He's a friend's father, I'd known him for years, and he explained, too, how those values of openness, freedom, and revolution had been part of the early computer era. They were why he made that rainbow logo. He no longer lived in San Francisco now but had been in the city not long ago for a meeting. He'd been summoned by the founder of a Japanese company, Crooz, because the man liked the Apple logo and wanted Rob to design one for his company. The CEO talked about his company's name and the waters surrounding us, in our bodies and the seas, and how much of the world was water. Crooz, which did everything from online gaming to e-commerce, wanted to be that water *and* the boat–to control where people were going. Rob's solution was a blue drop, flipped onto its side to make a C, with a round droplet at its center. He added one last thing about blue. "It veers to black," he said, "like gathering dusk outside." It was the same dark hue I'd seen in the color as I set out on my search, and maybe also the hue of Homer's wine-dark shade.

Apple's first logo wasn't, in fact, Rob's. For nearly a year the company had used something that looked like a Dürer etching. It was Newton reading under an apple tree as a piece of fruit threatened to fall on his head. A legend at the bottom declared, "Newton: A mind forever voyaging through strange seas... alone." Somehow that seemed fitting, tying back to water but also, in a sense, to color. In the 1660s Newton had been the first to refract light through a prism to get the spectrum. He identified it as having seven colors, including blue, indigo, and violet. Indigo was later dropped from the list. It was thought that he'd only included the shade to make up the number seven, which was rich with symbolism, and his indigo seemed indistinguishable from blue and violet. In fact, Newton's blue had probably been cyan.

Stewart Brand was another of the Berkeley hippies who saw possibility, peace, and freedom uniting in computers. He created the *Whole Earth Catalog* in 1968. Not a catalog at all, it didn't sell anything but instead reviewed tools and technology (though the definition of tool was broad) and ran articles and essays. Brand's guiding ethos was empowering people through information.

In the first edition, he wrote that governments and industries, with their top-down bureaucratic ways, had failed. Instead, he said, "A realm of intimate, personal power is developing–power of the individual to conduct his own education, find his own inspiration, shape his own environment, and share his adventure with whoever is interested." The *Whole Earth Catalog* aimed to provide what the internet would promise decades later, and Brand saw all the hope that free information could conjure. Nearly thirty years later, Steve Jobs compared the *Whole Earth Catalog* to Google in a commencement speech at Stanford. In the '60s Brand had also campaigned for NASA to release the first image of the earth from space: the whole earth, he'd called it. That image of the globe as a tiny marble appeared on the catalog's covers.

The same year the catalog launched, Brand attended what's now called the Mother of All Demos, where the first working prototype of a computer as we know it was displayed by Douglas Engelbart. It had a graphic interface, mouse, windows, hypertext, even videoconferencing. Brand and Engelbart ran in similar circles, and Brand even helped out at the launch. In 1985, before the first webpage had been created, Brand went on to cofound the WELL, one of the first virtual communities in the United States. Its members even coined the phrase "virtual community." Standing for Whole Earth 'Lectronic Link, WELL also had a utopian, hippie feel, as if it were embracing the future with free information and a communitarian spirit. I remember living in New York in the early '90s and being jealous of how the WELL created a sense of a shared new world for its members.

The logo was the name in all caps, except the *e*. It was lowercase, as Microsoft Explorer's would later be, with a swirling circle at the center like a whirlpool–or perhaps the globe. I thought now of how ideas could be linked to a place, as Lynn described. Here had been a new focus on the individual in the most optimistic sense. The "whole earth," the globe, the WELL, those images of water, those blues that shaped the early internet came from Brand. So, too, did the saying "information wants to be free." He said it in the mid-'80s, when the internet as we know it didn't exist, and when he said *free* he meant money. He was talking about how distributing information was getting cheaper and cheaper, and his original statement was much longer than the sound bite it's been reduced to. "Information wants to be expensive," he said, "because it's so valuable. The right information in the right place just changes your life. On the other hand, information wants to be free, because the cost of getting it out is getting lower and lower all the time. So you have these two fighting against each other."

By the early 2000s, the phrase had evolved into the clarion call for an open internet with universal access. This idea of free information is also one with no boundaries. It's easy to slip from openness to neoliberalism, though, where that goal of openness is misinterpreted into a laissez-faire system with no protections for users; where businesses, particularly ones online, can conflate the communitarian urge for an open and free internet available to all with the free, unregulated flow of money.[7] It's a quick slip of language for social-media moguls who want everyone to have access to their sites while maintaining no responsibilities for what that access entails–all the while searching, tracking, and selling users' data.

"I like technology that is unbiased," Jack Dorsey said. But technology has never been unbiased. Today Mosaic's and Netscape Navigator's creator, Marc Andreessen, embraces libertarian causes.[8] He is on the board of Facebook and runs a venture capital firm, while Mark Zuckerberg, who could see all of blue, campaigns for internet freedom, but the freedom he wants is one in which everyone has access to Facebook–and Facebook has access to all those users' information. This Berkeley hippie language is being used by the tech industry to posture as if it's creating a better, purer

---

7. These are conditions that have also made information valued less—just ask a writer how sites like Facebook and Medium aggregate news, writing, and user content. No one pays for news, so people devalue it. Voilà: the 2016 election. I came of age in traditional media, and outside its realms I have a chance to write essays like this. The internet has created spaces for more experimental writing. But money is still a question, and in that question of money is also one about who gets to write what. Whose writing are we reading, and whose news? A single mother, for example, likely can't support herself as an adjunct professor, so probably won't get to write essays like this.

8. His tilt towards libertarian beliefs is ironic given that his first major project, Mosaic, was funded by the federal government while he was attending a public university.

world. I wasted an entire afternoon counting Zuckerberg's posts mentioning freedom and openness online. "For the first time ever," he said in one, "one billion people used Facebook in a single day." In his Congressional testimony about Cambridge Analytica, he talked of his company's "social mission" and called Facebook "idealistic and optimistic." The information he collects–the data and metadata and digital signatures, the things we sign away knowingly and unknowingly–is hardly different from the information the CIA's drone operators use to target terrorism suspects. Only Facebook uses that information to target ads.[9]

Back at the military base-turned-national park where I had been writing for the past month, I hiked up a hill. It was on the edge of the ocean. Waves crashed in the distance, and there was no one nearby. This place was pristine because while it was a base there had been no commercial development here. I was leaving in a few days to see my sisters, and I stood on the bluff by myself and pondered the word *alone* in the first Apple logo with Newton. *Voyaging strange seas alone.* Apple had made Engelbart's designs accessible to millions. Today those innovations are integral to our experiences online. The sky was limpid blue all the way to the horizon, where sea and sky became one, and I kept thinking of language and what got secreted into it and how ideas might resonate over time, haunting our present. The screen had first been called a window in the late '60s, around the time DARPA developed the internet. Now that screen opened onto a new landscape. Yves Klein called his monochromes "landscapes of liberty." The sky, the window, hope... computers were filled with idealism. By the time Netscape Navigator launched, that freedom was tied to commerce, and Lynn saw very quickly that commerce would lead to our being tracked.

After Steve Jobs drained all the color from Rob's logo, he moved his company offshore to Ireland to avoid taxes, embracing that neoliberal ideal that money could travel without borders. The borderless, unchecked internet was the

---

9. Like algorithms, drones' remoteness works to absolve human involvement and responsibility for war, particularly now that AI analyzes images.

goal of people like Andreessen and Zuckerberg, and this internet celebrated individualism, even as Zuckerberg often called Facebook a community.

I hiked down the hill to a Nike nuclear missile site that was now a museum. Behind razor wire, the sharp white nose of a missile pointed into the air as if about to launch. After a season of El Niño, a winter of winds and rains, the hill was dotted with wildflowers, but once it had been guarded by attack dogs and men with rifles. Nike missiles were installed here in 1954 during the Cold War, and later DARPA would create the internet as a tool for that same war. In World War II we'd seen the results of totalitarianism, how it shackles the self to the state. Forced conformity had been the near-death of civilization as we knew it, and afterwards communism loomed large. Instead the West embraced the opposite of the communal. Over the next few decades, the individual–aloneness–had been raised up and had become the answer. The individual equaled freedom. It was a liberation ideology of the self.

The Cold War became the Vietnam War, and in 1968 Brand was still talking about the individual. In light of the failures of the state and corporations, he saw the individual as the path forward. Free self-expression was to be our salvation, except that, with time, freedom transformed into the glut of selfhood splayed out on social media, and all of it came dressed up in blue.

It's no surprise that Cambridge Analytica got its data from something clothed as a personality test. We're all posting our curated selves online–me included. It's the snapping turtle in my yard, the protests I go to, the art I see, and this small shining moment of joy when someone likes them: a heart on Instagram, thumbs up on Facebook. That quest for connection and approval is deeply human. Companies exploit that need and render it addictive, and while they might call themselves communities, these platforms prioritize the individual, that curated self, over the group. This is what we are sending out into the world, and by *we*, I mean the United States; I mean that small corner of the United States where companies like Facebook and Google are based.

The individual has long been an American ideal. It's the pioneer, the yeoman farmer, the entrepreneur, the pull-yourself-up-by-the-bootstraps narrative. It's Mark Zuckerberg working alone in his dorm room at night. It is also Donald

Trump. We lionize this. The core of capitalism is the individual; it's the basis of Adam Smith's liberalism, which was "liberal" in that it had no restrictions, believing the market would create them, believing that when everyone acted out of their own self-interest the appropriate level would be found.

As companies have hijacked the idea of a free internet–"free" only in that it shucks boundaries and regulations–those companies that thrive on advertising dollars get more money. *More than a billion a day on Facebook*, Zuckerberg wrote. All of this is blue. The blue is freedom, expression, the air, the clouds, the sky, the window onto the world. It is innocuous. It seems unthreatening. It is a "landscape of liberty," and it is trust and openness. It is also privatization, the good of the individual over the group, rather than a greater civic good.

Overhead a hawk circled and screeched. I watched it grab a snake and fly off, the long limp body in its talons. I stood at the missile site's gate and thought about my dad writing his letter to Adlai Stevenson in 1952. He'd spend that next year driving hundreds of miles on his own time in his own car to Albany for hearings, testifying and nagging and lobbying various politicians about protecting public power projects, so that private companies couldn't exploit public resources. Then he spent the Cold War building rural electric cooperatives in parts of the country that no utility company wanted to serve because they were too poor and too rural, with too small a population for there to be anything to gain for the utilities. He believed in collective ownership. He would have seen the internet in this way, too, something requiring universal access–collective ownership. I thought also about Stewart Brand, who believed in shared information, and the utopian hopes he'd had in the WELL and in the *Whole Earth Catalog*.

I walked down the bluff to my studio and emailed Brand. His office was nearby, and Lynn suggested I get in touch with him about meeting. "Thanks," he responded almost immediately, "I have to pass." Five brief words, the text highlighted in blue on my screen.

Into this blue go all the data points of our lives. They get mined and harvested by algorithms. There's little difference between how data is collected in contemporary warfare and how it's collected in ad targeting. Data is still being used to analyze, control, and, in the case of drones, kill with inhuman results–in the lineage of McNamara. The flipside of all this freedom and unbridled selfhood is something totalitarian. They're entwined. This is what blue conceals. It looks like hope. It comes from those hippies in the Bay Area, from Stewart Brand with that marble of the globe, the one Microsoft stole. The blues evolved from something optimistic to something that camouflages surveillance. This is what I found in blue.

I stand at the edge of the ocean one last time before I leave. In this national park that had been a military base, I feel like I've reached the edge of the earth. I am about to return home, and I want this blue to be hopeful. It is a tiny, postage stamp-sized painting and a representation of the sea and sky–and the dream of unlimited freedom. ●

'I have nothing to hide' is another way of saying 'I have privilege.'

Alvaro Bedoya talks with Cindy Cohn

O ne of the often overlooked places where we can glimpse the end of trust is in the government's treatment of us, its people. Over the last eighteen or so years, the United States government has largely abandoned the idea that its people can associate, communicate, or even show their faces in public without the government having access to that information. Instead, locally and nationally, we see a series of government tactics—facial recognition, license plate readers, access to the internet backbone, searches of the enormous databases of tech giants, real-time access to private security-camera footage—that widely, and largely indiscriminately, subject our lives to governmental surveillance.

When these techniques are considered in Congress or in broader public debate, people often reference 1984 or other cultural touchstones where everyone is equally watched. Yet the lived reality is that the effects of these techniques are not equally distributed throughout society. Surveillance, even mass surveillance, has been and continues to be disproportionately aimed at people of color, immigrants, and other marginalized members of society. It also disproportionately impacts those who are engaged in politics or activism, who are often the same people.

I talked with Alvaro Bedoya, Executive Director at the Center on Privacy & Technology at Georgetown Law, about this disparate impact and about how any conversation on privacy is incomplete without recognizing this reality. We also talked about some ways forward under the Constitution, and the fact that Harriet Tubman was basically unstoppable.

CINDY COHN: I want to talk a bit today about the disparate impact mass surveillance programs have on people of color—programs like tapping into the internet backbone, gathering up telephone records, scooping up license plates, and utilizing facial recognition software. I know you've thought about this a lot, Alvaro, so tell me how you think about these things.

ALVARO BEDOYA: So let me tell you about my grandmother, my Abuelita Evita.

My grandmother was from a town in the Peruvian highlands called Cajamarca. She was born in 1914. She grew up in an old house with a Spanish courtyard surrounded by centenarians. Or actually, people who lived to be ninety-nine– not a hundred, not ninety-eight–ninety-nine. *She* lived to be ninety-nine, her mom lived to be ninety-nine, and her aunt did, too. Evita became a school-teacher and then eventually principal of an elementary school. Because she was a schoolteacher for forty or fifty years, she had taught pretty much half of the town by the time she retired. She had a pet chicken, didn't trust banks. I remember going to her house when I was little and discovering wads of this defunct Peruvian currency–I think it was *intis*–underneath frying pans in the kitchen. She was straight out of a Gabriel García Márquez novel.

Anyways, after my family came to the U.S. in 1987, every weekend, no matter how much the call cost, we would call and check on Abuelita Evita.

So–why am I talking about her? Because in all likelihood, from the 1990s to 2013, every single time we called Evita, a record was created by the Drug Enforcement Administration in what is now recognized as the precursor to the NSA call records program. This program basically tracked all calls from the U.S. to a large number of countries abroad, eventually including most of Latin America.

It wasn't just me. It was likely every other immigrant calling their family in Latin America in that period. And it didn't happen to people who were calling their grandmother in Topeka or New York or wherever. It happened to immigrants calling their family abroad, especially in Latin America.

I think there's a problem when people talk about surveillance and focus on the "mass" part of surveillance, the "everyone is watched" part of surveil-lance. When we focus only on the fact that "everyone" is watched, you lose the fact that some people are watched an awful lot more than others. And you also lose a sense of what the value of privacy is to the particular communities affected most by its absence.

The NSA call records program, our nation's broadest and most invasive admitted domestic surveillance program, was arguably beta tested on Amer-ican immigrants. And that's part of a pattern. But if you listen to the way elected officials, or even the news media, talk about surveillance, they don't talk about that.

CC: That's really right. When we talk about mass surveillance, it's easy to miss that we're talking about techniques that have two fundamental steps: first, a look at everything—whether that's over an internet juncture or via a passing license plate reader or through a policeman's body camera. That's bad enough and itself can disproportionately impact marginalized people since we know they are over-policed. This was true for the DEA as well, when it targeted Latin American immigrants calling their grandmothers but not Iowans like me.

But it's worse. There's then a second step, when the government's "targets" are pulled out of the mass stream. In the NSA cases, that list is claimed to be a state secret. And for the DEA, we still don't really know. So at the first step it's bulk or mass surveillance that can be disproportionately targeted. And when you get to the second, there are also people who get affected more than others. At both steps those affected are disproportionately people of color or other marginalized groups, including, of course, immigrants. At the end of the day there are people whose real lives are affected by this more than others and to elide that is to miss a piece of what's really going on.

AB: People who work on these issues often use a worst-case scenario where the United States becomes a dystopian surveillance society and everyone is a suspect. Well, I think there're lots of communities that *already* live in a surveillance society. A lot of working-class, primarily African American urban communities already live in a surveillance society. Look at Baltimore. We have seen persistent aerial surveillance, the use of these devices called Stingrays that mimic cell towers, face recognition—you name it.

The same is true of immigrants who are undocumented or who live in mixed-status families. You see a lot of these technologies that were first developed for the battlefields of Iraq and Afghanistan being used most aggressively to find these immigrants. But when you see "surveillance" talked about in the press, it's typically not used to refer to those folks. It's used to talk about "everyone."

And by the way, this isn't a new thing. This is a very old thing. If you think of a civil rights leader, an African American civil rights leader from the twentieth century, the chances are strong that they were surveilled in the name of national security. It's not just Martin Luther King: it's Whitney Young, Fannie Lou Hamer, W.E.B. Du Bois, Marcus Garvey, Muhammad Ali, Malcolm X. The list goes on and on and on.

CC: The FBI's COINTELPRO program that ran from 1956 to 1971 surveilled the Puerto Rican Independence movement, the Black Panther Party, Cuban and even Irish Nationalists, along with the anti-Vietnam War activists, the Communist Party, and the Socialist Workers Party. And that's just what we know about.

What's new is that modern surveillance techniques can scoop up a wider and wider range of information during that step one I mentioned. The government can also now quickly sift through and store more data than ever before.

That means that unlike all of those other situations, it's a lot harder now to hide in the crowd. And it means that we're at risk of losing lots of things that get done in that relative anonymity of the crowd.

The experiment of the United States is an experiment in self-government, but inherent in it is the idea that the people can change their government. You need to be able to have a conversation that the current leaders of the government don't want you to have–about how you want to vote them out of office or change an unjust law or a law being applied unjustly. I think we're building a technological world where we're going to have to be more explicit about the importance of that private conversation. When the country was just getting started, it was technologically harder for a government to know what everyone–or even specific people–were saying to each other, so implicitly everyone had the ability to have a private conversation.

AB: Look at facial recognition technology. This is technology that lets law enforcement point a camera at a group of protesters, scan those people's faces, and pick out, by name, who's standing in that crowd. This was used in Russia to name and publicly shame anti-Putin protesters, supporters of Alexei Navalny. Here in the states, in Baltimore, you saw it used to identify and arrest people protesting the death of Freddie Gray. Police took photos that had been posted on social media, scanned those faces, found people who had outstanding arrest warrants, and then arrested them. This is real.

Privacy protects different things for different people. If I'm struggling with mental health, maybe it allows me to get help or just lets me know that I'm not alone. If I'm gay and growing up in a conservative family, maybe it lets me explore my sexuality online. For certain communities, privacy is what lets them *survive*. Privacy is what lets them do what's right when what's right is illegal.

Take Harriet Tubman. What she was doing–freeing enslaved people–was very much illegal. And not just in the Confederacy; it was very much illegal in the United States before the Civil War. For Harriet Tubman, it was her ability to evade capture, to evade detection, that let her do the things that we celebrate her for today. It's what let her repeatedly go to the same area, the same plantations–the places where she was *most* hunted and wanted–and continue to free people. She did it over and over again. Before we build a system that is literally capable of tracking everyone and everything, I think we need to ask ourselves if we're building a world without underground railroads.

CC: Harriet Tubman was such a force. I wonder if we're preventing the forces of the future here.

One of the things as a constitutional lawyer that I take a little heart from, though, is that underneath the technological changes, this is a problem that the framers of the Constitution and Supreme Courts of the past actually understood pretty well. They understood the need for people to have private conversations because, frankly, that's how our country came into being. So if we can frame things correctly, there are some parts of the Constitution that can help us here.

For example, the First Amendment doesn't just protect the right of free speech–it protects the right of assembly, and the right of association. To me the right of association is especially important for us to talk about in relation to people of color because it was used in exactly this way during the civil rights era. There are a series of lawsuits that arise out of the NAACP's organizing in the South. The leading one is called *NAACP v. Alabama*, where the state of Alabama was requiring that the NAACP hand over its membership lists as a condition of operating in the state. The Supreme Court struck down that requirement as violating the First Amendment right of association. The logic was pretty simple: the court recognized that ordinary Alabamans would be afraid to join the NAACP if they knew that the state of Alabama would be informed. Remember, we're talking about the Alabama of Governor George Wallace here: extremely hostile to civil rights and the NAACP. The Supreme Court said of course it's not workable to require an organization like the NAACP to have to hand over its list of associates to the government. This rule is necessary in a society that embraces the idea that it's the people who have control of their government, not the other way around.

So what's actually happening is our government is using our "metadata"– who we talk to, when and how often and from what location, and using it to map people's associations. They call it "contact chaining," which sure sounds nonthreatening. But if you drop the terminology and talk about the fact that the government is claiming the right to track our associations, then the *NAACP v. Alabama* case can help us argue that the First Amendment requires the government to meet a very high burden, something we call "strict scrutiny," before it can do so. This requires the government to have both a very important goal and a tight fit between the technique used–contact chaining–and the goal. If strict scrutiny applies, the government should not be able to collect this sort of mass data since it impacts large numbers of non-suspect people. The fit isn't tight enough.

That's one of my major goals for the coming years–to help Americans, including American judges, see that digital tracking of people's associations requires the same sort of careful First Amendment analysis that collecting NAACP membership lists in Alabama did in the 1950s. Use of license plate readers to track people going to church or to the mosque, or to track the calls of little boys to their grandmothers in Peru, requires this careful analysis regardless of the use of sanitized words like "metadata analysis" and "contact chaining." But the good news is that for those of us who have to go in and convince judges of things, there are strong bases in the Constitution for this concern.

AB: There are. That said, there is a lot of work to be done in this respect because, for example, the *NAACP* case had to do with preexisting membership lists that I think the government was trying to get from the NAACP. Nowadays, with face recognition, law enforcement can actually *create* those lists by photographing and scanning people's faces. Recently, the Supreme Court ruled that we do have a right to privacy in our public movements for periods longer than a certain interval. The police can't track you for a week without getting a warrant. But we don't know how that ruling will apply to, say, police taking a snapshot of people at a protest and then scanning their faces to identify them.

I share some of your optimism, though, because I do think that people get it. Sometimes, people don't get upset about cookies or other online digital tracking methods because it's not *real* to them. But a technology like face recognition, they get that. They get that giving the government the ability to track your face is invasive and just creepy, frankly.

It's on people to come to hold their government to account and say, "Hey, just because you could do something in the past that is remotely related to the thing that you're now doing to track us doesn't mean you can do it now. It's different." So, for example, there are old court cases saying that it's okay to photograph protesters. But face recognition is different from photography. People need to argue that and recognize that.

So I do think there's hope, but at the same time, unless people act on that hope there's a real risk of doing away with some of our oldest protections. For some people, that's a question of civil liberties. But for others, it's a question of survival.

CC: I think that there is a way in which things that are not visible to people who are in the majority culture are very, very present for people who are not, and bridging that gap is really important. There's that whole old trope that "if you don't have anything to hide, you don't have anything to fear from surveillance." I often try to get people in the majority to step back and think not just about themselves but everybody who they know, everybody who they might be related to, and even more broadly what kind of society they want to live in. Do you want to live in a society where the only government you ever get is the one that you have now? Every social movement had to start with a private conversation. Abolitionists and suffragettes were not always safe to meet in public; abortions and even some adoptions happened through whisper networks. In my lifetime talking about gay people getting married was something that could get you killed. So this isn't just about the past.

But every social movement we celebrate today started in a period where conversations had to be had in secret, or at least without the current government listening in, before they could organize enough and gather enough people together to change things.

I also think that, looking at how social change happens, you need systems that not only protect the few brave people who are willing to go to jail for their beliefs–that's really important–but you also need systems that ensure that it's safe for many more people to show their support. That includes people who are busy raising their kids or working two jobs or who are more vulnerable in other ways. I worry that sometimes when we focus just on the big-name heroes of our movements, we miss the need to protect the entire pathway from the few to the many. Without that pathway it's not possible to move society, whether the movement comes through passing laws, winning elections, or securing court victories.

AB: I think that "I have nothing to hide" is another way of saying "I have privilege," or "I'm a relatively powerful person who is from the right side of the tracks, who has political opinions that aren't considered radical, who has the luxury of being the right gender and sexual orientation." We need to stop talking about privacy as this vague, undefined thing. We need to recognize that it is a shield for the vulnerable. Privacy is what lets those first, "dangerous" conversations happen.

CC: Another thing I wanted to think about with you is this idea that privacy is more of a team thing, and how, especially when we think about mass surveillance, we often forget that. Instead, it all comes down to "Do I have something to hide?" "What about me?"

But focusing on individuals, and specifically only ourselves as individuals, is really missing a lot of the mechanisms that the Constitution is actually thinking about with the right of association, assembly, to petition for grievances.

Yet what those of us working on these issues get told by the marketing professionals is that unless you can make it personal, the people are not going to care about it. I think one of the fundamental things that I struggle with is how to talk about something that is really about the collective in a society where all of the marketing and hoopla is about me, me, me. For many people in the majority culture, they are right that they might not ever be the target of surveillance, and that while they are subject to the mass surveillance techniques–the license plate reader, internet backbone surveillance, facial recognition cameras–the results of those techniques may never be turned against them.

AB: The law struggles with privacy as a communal right. For example, in 1967, when the Supreme Court set out a test to figure out the definition of "search" in the Fourth Amendment, it decided that the test included a "reasonable expectation of privacy." It asks two questions: first, whether you have an expectation of privacy personally, and second, is that an expectation that society is prepared to recognize as reasonable? It's a really weird test if you think about it because, under that test, the government could go around town plastering posters saying "We're watching everyone and hearing everything"– and then, in theory, no one would have an expectation of privacy and the government could search everything.

But the Fourth Amendment should not be defeated so easily. It's about more than my individual right to keep the government out of my business. It's also about protecting a certain kind of relationship between the government and its people.

How do you get enough people standing up for that communal right to privacy? That's the billion-dollar question. I don't know if us lawyers are going to solve that. I think that that question will be solved by artists and writers and people involved in our culture, rather than me, a lawyer who loves to nerd out on obscure precedents.

cc: Yeah, I think that's really important. Though I do want to throw out there that, in terms of those things the Fourth Amendment doesn't do very well, the First Amendment *is* actually *trying* to do some of them. It's a tool that we ought to be able to use a bit more. For instance, in the device border search case that EFF and ACLU are handling called *Alasaad*, the court recently recognized that searching people's devices at the border, especially journalists, is a First Amendment problem because the searches reveal all the people they talked to.

Part of why we're having this conversation in *McSweeney's* is to try to reach out to the artists, writers, and our culture more broadly. But I do also think that there are tools on the legal side. The good news, since I want to end on a good note, is that I actually think there are parts of the law that are there waiting when we're ready as a society to talk about these things.

AB: Amen. Let's end on a good note. Not a crazy depressing one. ●

A COMPENDIUM OF LAW ENFORCEMENT SURVEILLANCE TOOLS

By Edward F. Loomis

○ Facial Recognition Systems

○ GPS Trackers

○ License Plate Readers

● **Drones**

○ Body Cameras

○ Cell Tower Simulators

○ Parallel Construction

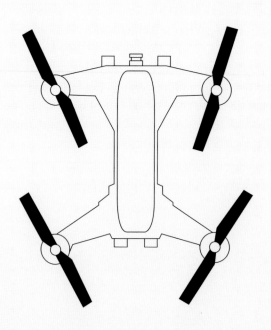

Many federal, state, and local police departments now employ drones for covert surveillance, crowd monitoring, crime scene analysis, investigating active crimes, reconstructing traffic collisions, and search and rescue operations. The Center for the Study of the Drone at Bard College issued an updated drone usage report in May 2018 that indicates 302 county police and sheriff and 278 municipal police departments currently own at least one drone, and that the number of public safety agencies with drones had increased by approximately 82 percent in the last year alone. It reported the largest local law enforcement drone inventories to be owned by the Polk County, Fla., Sheriff (20), the San Diego County Sheriff (10), the Alameda, Calif., County Sheriff (10), the Palm Beach, Fla., County Sheriff (9), and the Rockford, Ill., Police Department (9).

Drone technology has advanced considerably in recent years and some drones can conduct surveillance operations from heights of four miles—far enough to be invisible from the ground. When equipped with high-resolution cameras, drones can stream live video and generate massive amounts of data. More sophisticated drones can be armed to penetrate wi-fi networks and record phone conversations and text messages. They can also carry infrared sensors that monitor persons at night and inside buildings.

Reuters reported on August 2, 2017, that the Secret Service planned to employ a drone to monitor President Trump's Bedminster National Golf Club for an upcoming visit. The Secret Service stated it would notify local residents that the outer perimeter would be monitored and that any recordings made would "be overwritten within 30 days or become part of a law enforcement investigation." It's yet to be determined what the legal implications of surveilling the general public with this technology are regarding Fourth Amendment privacy rights.

Accounts of U.S. predator drone strikes against enemy combatants outside of war zones date back to their use against Yemeni terrorists in November 2002. Although no reports of police department use of such lethal technology have appeared, former Attorney General Eric Holder confirmed in his May 22, 2013, letter to the Senate Judiciary Committee that four American citizens had been killed by "U.S. counterterrorism operations against al-Qaida and its associated forces outside of areas of active hostilities." The specific language was intended to avoid identifying the means as drones. ●

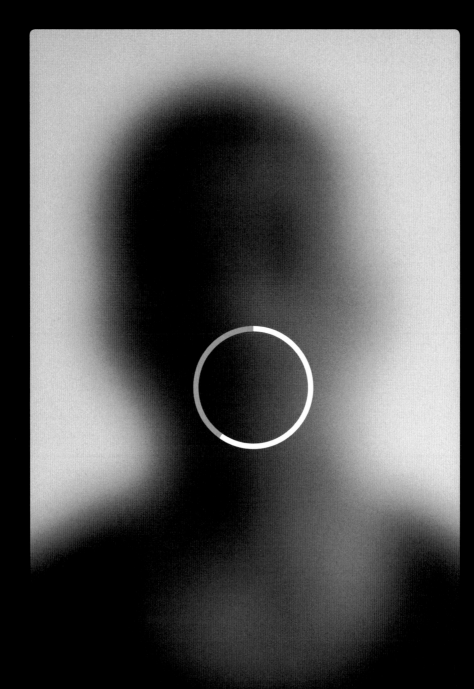

Reconsidering Anonymity in the Age of Narcissism

Gabriella Coleman

To all the names in history: The time has come to sacrifice that name.
                              –Anonymous (*Manifesto of the Anonymous Nomad*)

Ego & fame are by default, inherently contradictory to anonymity. The tallest blade of grass gets cut first. Remain unknown. Be #Anonymous.
                    –Anonymous (@YourAnonNews), April 16, 2012

The premise of this collection is that privacy and anonymity are vanishing under the onslaught of government and corporate surveillance. The premise is not a new one; in 2009 many advocates, activists, librarians, and civil libertarians were finding it impossible to imagine privacy and anonymity existing into the near future. This was a time when Silicon Valley executives were building the digital infrastructure of surveillance capitalism and defending it by casting privacy as morally dubious. For instance, when Google's Eric Schmidt was asked by a reporter whether we should entrust our data to them, his patronizing response was calculated to eliminate any positive valence to privacy: "If you have something that you don't want anyone to know, maybe you shouldn't be doing it in the first place."

But around that same time a mysterious collective bearing the name Anonymous came to prominence–a far-flung global protest movement predicated on the idea that cloaked identities could be put to work fighting for justice by enabling truth-telling and disabling celebrity-seeking behaviors. While initially used by nameless trolls coordinating one-off harassment escapades across the internet, the Anonymous moniker took on new meaning in 2008, as participants identifying with the label engaged in a staggering array of hacks and political operations designed for media uptake. Figures identifying as Anonymous used their technical know-how and trollish sense of media spectacle to call for a moratorium on Japanese and Norwegian whaling; demand justice for victims of sexual assault and police brutality, sometimes by revealing the names of alleged perpetrators; hack governments and corporations alike; assist the occupations in Egypt, Tunisia, Spain, and North America; support

the Syrian uprising; dox police officers who pepper-sprayed protesters; expose
pedophiles online; and even provide clothing to the homeless. News outlets
came to count on Anonymous for a steady stream of sensational stories. One
affiliated crew called LulzSec devoted itself to delivering a new "hack-a-day"
for fifty days. As they infiltrated Sony Pictures, published fake news on PBS's
website, and snatched emails from the Arizona Public Safety organization,
they served up fodder to the press even as they gleefully self-reported their
exploits on social media to a growing and satisfied fan base. "In the last few
weeks these guys have picked up around 96,000 Twitter followers. That's
20,000 more than when I looked yesterday. Twitter has given LulzSec a stage
to show off on, and showing off they are," wrote one security researcher.
Anonymous managed to court even more controversy with ritualized stunts
like "FUCK FBI FRIDAY," which saw the hacktivists take to Twitter at the end
of each week and taunt the agency tasked with snuffing its members out. For
an anthropologist who studies the cultures of hacking and technology, it was
an exhilarating moment; I was glued to my seat.

But as that exemplary moment passed, the story of Anonymous veered
towards the ironic, and ultimately even tragic, as the core participants were
betrayed and arrested, and the name began to lend itself to military operations–such as anti-terrorism campaigns in service of the nation-state–that
many of its earlier members would have at times vehemently opposed. Given
the omnivorous power of the contemporary digital surveillance machine to
coax data from humans and then use it against us, I was never so naive as to
actually believe that Anonymous could be our saviors. My take was humbler: I
mostly marveled at the way these masked dissenters embraced anonymity as
a sort of ethic to prevent social peacocking behaviors and to motivate participants into silent solidarity rather than individual credit-seeking, even as they
were hounded, and sought collective publicity, for their epic hacks, pranks,
and protests. It certainly helped that Anonymous contributed to a number
of political causes I supported, such as Occupy Wall Street, the exposure of
surveillance firms, and struggles against government corruption. I appreci-
ated that groups of people were taking up the mantle of anonymity largely

for good–even if it seemed it might be for one last time before anonymity itself dissipated altogether.

My pessimism about the viability of anonymity and privacy to survive (much less thrive) still generally overpowers my optimism. But even as the glory days of Anonymous waned, a slightly more muscular privacy and anonymity movement finally coalesced. Thanks in part to Edward Snowden's massive leak of NSA documents, which provided much stronger proof of government surveillance and its collusion with the private sector than had previously existed, a battle to preserve privacy and anonymity is now being vigorously waged. Shortly after the Snowden disclosures, countless hacker-driven technology projects, galvanized by his exposé, continue to develop the sort of privacy-enhancing tools that journalists, domestic-violence victims, human-rights workers, and political dissidents now rely on to move through the world more securely. The usability of these tools has considerably improved. Whereas five years ago I struggled to recommend simple security tools to friends and family, today I can point to Signal (an encrypted texting and phone application), the Tor browser (which anonymizes web traffic), and half a dozen other applications, each of which has garnered increased funding and volunteers thanks to increased scrutiny of state and corporate privacy violations. Even Google announced that they would instantiate strict end-to-end encryption of its services to ensure the data it relies on to fuel its commercial enterprise would not be so easily available to others, though they've yet to carry out these changes. Existing policy, technology, and advocacy organizations like the Electronic Frontier Foundation, Fight for the Future, the Library Freedom Project, Big Brother Watch, and Privacy International have also helped ensure that privacy remains a marquee political issue. A steady stream of new scandals, such as the revelations that Cambridge Analytica used personal data harvested from Facebook to influence election results, has amplified these concerns, and demonstrated the extent to which questions about personal data and privacy remain very much unsettled.

As a member of a loose confederacy of anonymity-defenders, I routinely give lectures about the ways anonymity can enable democratic processes like

% of Internet Users Who Have Experienced Security-Related Issues

21%

▲ Have had an email or social-media account compromised

12%

▲ Have been stalked or harassed online

11%

▲ Have had important personal information stolen (*i.e.* SSN)

6%

▲ Have lost money as a result of an online scam

6%

▲ Have had their reputation damaged because of an online event

4%

▲ Have been led into physical danger because of an online event

Source: Pew Research Center

dissent and whistleblowing. In the course of this proselytizing, it has become apparent that anonymity is often harder to defend than other closely related civil liberties like free speech and privacy. Anonymity gets a bad rap. And it's not difficult to see why: the most visible uses of anonymity online, like comments forums, tend towards the toxic. Numerous newspapers in recent years have eliminated these forums, reined them in, or reconfigured them, attentive to the ways they often fail to engender civil discourse and instead breed more hateful and harmful speech. Anonymity similarly enables trolls on social media to dodge accountability as they viciously attack (mostly) people of color, women, and the genderqueer.

The negative connotations that many have of anonymity is evident in their perception of what journalists and scaremongers call the dark web. When I ask my students what they think happens there, many describe it as the most sinister corner of the net, infested by menacing pervy types who hack bile onto our devices, festering and erupting into mini-volcanoes of stolen passports, cocaine, and child porn. Some even believe that being anonymous online is tantamount–in every instance–to trawling the dark web. The metaphor of darkness has clearly worked to implant nefarious and inaccurate pictures in their minds, so I counter with a different image.

Since my students have little understanding of how anonymity works, first I explain that, far from being a binary choice like a light switch that turns off and on, anonymity typically involves an assortment of options and gradients. Many people conceal themselves by name alone, contributing online with a screen name, alias, nickname, avatar, or no attribution at all: "anonymous." This social anonymity concerns public attribution alone and shields a participant's legal name, while identifying information, like an IP address, may still be visible to a network observer such as the system administrator running the site where content is posted. There is also no single godlike anonymity tool providing omnipotent, unerring, dependable, goof-proof protection with the capacity to hide every digital track, scramble all network traffic, and envelop all content into a shell of encryption. Far from it: flawless technical anonymity is considered a demanding and exacting art that can occasion loss of sleep for even the most elite hackers. A user seeking out technical anonymity must patch together an assortment of tools, and the end result will be a more or less sturdy quilt of protection determined by the tools and the skill of the user. Depending on which and how many tools are used, this quilt of protection might conceal all identifying information, or just some essential elements: the content of exchanged messages, an originating IP address, web browser searches, or the location of a server.

The same anonymity, I continue, used by the criminal or bully or harasser is also a "weapon of the weak," relied on by ordinary people, whistleblowers, victims of abuse, and activists to express controversial political opinions, share sensitive information, organize themselves, provide armor against state repression, and build sanctuaries of support. Fortunately, there is no shortage of examples illuminating the benefits derived from the protection of anonymity: patients, parents, and survivors gather on internet forums like DC Urban Moms and Dads to discuss sensitive topics using aliases, allowing for frank discussions of what might otherwise be stigmatizing subjects. Domestic-abuse victims, spied on by their perpetrators, can technically cover their digital tracks and search for information about shelters with the Tor browser. Whistleblowers are empowered today to protect themselves like never before given the availability of digital dropboxes such as SecureDrop, located on what are called onion, or hidden, servers. These drop-off points, which facilitate the anonymous sharing of information, are now hosted by dozens of established journalism venues, from the *Guardian* to the *Washington Post*. Hosting data on onion servers accessible only via Tor is an effective mechanism to counter state-sponsored repression and censorship. For example, Iranian activists critical of the government shielded their databases by making them available only as onion services. This architecture makes it so the government can seize the publicly known web server, but cannot find the server providing the content from the database. When the web servers are disposable, the content is protected, and the site with information directed at empowering activists can reappear online quickly, forcing would-be government censors instead to play a game of whack-a-mole. Relying on a suite of anonymity technologies, hacktivists can safely ferret out politically consequential information by transforming themselves into untraceable ghosts: for example, one group anonymously infiltrated white-supremacist chat rooms after the tragic murder of Heather Heyer and swiped the logs detailing the workings of hate groups organizing for the Charlottesville rally, as well as their vile reactions and infighting.

Still, it is true that terrible things can be accomplished under the cover of technical anonymity. But it is necessary to remember that the state is endowed with a mandate and is significantly resourced to hunt down criminals, including those emboldened by invisibility. For instance, in 2018 the FBI requested around 21.6 million of its $8 billion annual budget for its Going Dark program, used to "develop and acquire tools for electronic device analysis, cryptanalytic capability, and forensic tools." The FBI can develop or pay for pricey software exploits or hacking tools, which they've used to infiltrate and take over child porn sites, as they did in 2015 with a site called Playpen. Certainly, the state should have the ability to fight criminals. But if it is provided with unrestricted surveillance capabilities as part of that mission, citizens will lose the capacity to be anonymous and the government will creep into fascism, which is its own type of criminality. Activists, on the other hand, who are largely resource-poor, are often targeted unfairly by state actors and therefore require anonymity. Indeed, anonymity allows activists, sources, and journalists not yet targeted by the state to speak and organize, as is their right, without interference.

The importance, uses, and meaning of anonymity within an activist entity like Anonymous is less straightforward than my earlier examples. This might partly stem from the fact that Anonymous is confusing. The name is a shared alias that is free for the taking by anyone, what Marco Deseriis defines as an "improper name." Radically available to everyone, such a label comes endowed with a built-in susceptibility to adoption, circulation, and mutation. The public was often unaware of who Anonymous were, how they worked, and how to reconcile their distinct operations and tactics. There were hundreds of operations that had no relation to each other and were often ideologically out of alignment with each other—some firmly in support of liberal democracy, others seeking to destroy the liberal state in favor of anarchist forms of governance. It's for this reason also that "Anonymous is not unanimous" became a popular quip among participants, reminding onlookers of the group's decentralized, leaderless character and signaling the existence of disagreements over tactics and political beliefs.

For members of the public, as well as my students, their assessment of Anonymous often depended on their reaction to any one of the hundreds of operations they might have come across, their perception of the Guy Fawkes figure, and other idiosyncrasies like their take on vigilante justice or direct action. While some spectators adored their willingness to actually stick it to the man, others were horrified by their readiness to break the law with such impunity. Amid a cacophony of positions on Anonymous, I invariably encountered one category of person loath to endorse Anonymous: the lawful good type (academic law professors or liberal policy wonks, for instance), always skeptical and dismayed at the entirety of Anonymous because of a small number of vigilante justice operations carried out under its mantle. The strange thing was the way those lawful types found agreement with a smaller, but nevertheless vocal, class of left activists—those keen to support direct action maneuvers but full of reservations when they were carried out anonymously. They tended to agree on one particular belief: that people who embrace anonymity for the purposes of acting (and not simply speaking), especially when such actions skirt due process, are by default shady characters because anonymity tends to nullify accountability and thus responsibility; that the mask is itself a kind of incarnated lie, sheltering cowards who simply cannot be trusted and who are not accountable to the communities they serve.

But these arguments ignore the varied and righteous uses of anonymity that Anonymous put in service of truth-telling and social leveling. With the distance afforded by time, my conviction that Anonymous has generally been a trustworthy force in the world and commendable ambassador for anonymity is even stronger today. Even if their presence has waned, they've left behind a series of lessons about the importance of anonymity that are as vital to heed as ever in the age of Trump. Of these lessons, I'll consider here the limits of transparency for combating misinformation and anonymity's capacity to protect truth-tellers, as well as its ability to minimize the harms of unbridled celebrity.

LESSON 1: TRANSPARENCY IS NOT A PANACEA FOR MISINFORMATION

Let's first consider the power of Anonymous and anonymity in light of the contemporary political climate, with journalists, commentators, and activists in a turbulent existential crisis over trust, truth, and junk news. Let me state from the outset that demanding transparency, in my political playbook, sits high on the list of expedient tactics that can help embolden democratic pursuits. Seeking transparency from people, corporations, and institutions that may have something bad to hide, and the clout to hide it, has worked in countless circumstances to shame con men and scumbags out of their coveted positions of power (and I resolutely defend anonymity for its ability to engender transparency). Still, the effectiveness of demanding transparency and truth has often been overstated, and its advocates sometimes naively attribute an almost magical faith to such a tactic while deeming the anonymous means to those same ends of truth-telling immoral. In the past, when I've discussed the importance of anonymity and the limits of demanding transparency in the pursuit of truth, very few people took me all that seriously besides a small group of scholars and activists already invested in making similar claims. All this changed when Donald Trump became president. Suddenly it was a lot easier to illustrate the logic behind Mark Twain's famous quip: "Truth is mighty and will prevail. There is nothing wrong with this, except that it ain't so."

Journalistic common sense, still largely intact leading up to the election, dictated that refuting falsehoods would preserve the integrity of the marketplace of ideas–the arena where truth, given enough airtime, can blot out lies. After Trump clinched the election, though, many journalists were forced to confront the fact that common sense, as anthropologist Clifford Geertz so astutely put it, is "what the mind filled with presuppositions... concludes." For critics, Trump's moral failings are self-evident in his dastardly behavior and pathological lying, both of which have been recorded meticulously by journalists. *The Washington Post* has tracked Trump's false or misleading statements since his first day in office, and found that his zeal for fibbing has only ballooned with time. However, though his supporters also discern Trump as audacious, they're armed with a different set of presuppositions and therefore reach radically different conclusions about his character and actions. In the same *Washington Post* audit of Trump's false statements, one online commenter shows how some of his defenders are willing to overlook his lies, interpreting him as authentic and emotionally forthcoming compared with the typical politician: "Trump is often hyperbolic and wears his feelings on his sleeve for all to see, refreshing some might say. One often wonders if it's even possible for him to be as duplicitous as the typical politician. His heart and policies do seem to be in the right place."

Appealing to those who distrust the contemporary political milieu, some of Trump's staunchest supporters argue that he serves a higher, nobler purpose by shaking up the establishment. Even as common sense can "vary dramatically from one person to the next," as Geertz put it, Trump has still managed to sequester our collective attention, baiting the media to cover his every move, often through a false yet convincing performance of authenticity. Whether in horror, amusement, or adulation, the American public stands together, beer in one hand, BBQ tongs in the other, mouths agape, mesmerized by his outrageously cocky antics. While some see the Trump presidency as an ungovernable slow-moving train wreck unfolding right before their eyes, others are clearly elated, cheering Trump on as if attending a monster truck rally. Trump is such an effective performer that he has not only managed to dodge any repercussions for his disturbingly brazen lying thus far, but also stands ready to accuse the establishment media of being liars: "I call my own shots, largely based on an accumulation of data, and everyone knows it. Some FAKE NEWS media, in order to marginalize, lies!" Under such a ruthless assault, truth struggles to prevail.

In contrast to Trump, Anonymous–a sprawling, semi-chaotic (though also fairly organized at times) string of collectives, composed of thousands of people and dozens of distinct groups acting in all four corners of the globe under its name, with loose to no coordination between many of them–comes across, in almost every regard, as a more earnest and trustworthy entity. While Trump helps us see this afresh, I've long made the following point: if one takes stock of the great majority of their operations after 2010, Anonymous generally followed a number of rather conventional scripts based on a drive to tell the truth. Anonymous would often pair an announcement about some indignation they sought to publicize with verifiable documents or other material. Such was the situation when Anonymous launched #OpTunisia in January 2011 and were some of the first outsiders to access and broadly showcase the protest videos being generated on the ground–footage they posted online to arouse public sympathy and spur media coverage. Anonymous routinely acquired emails and documents (and have, by the way, never been found to have doctored them) and published them online, allowing journalists to subsequently mine them for their investigations. Their drive to get the truth out there was also aided by splashy material engineered to go viral. Truth-telling, after all, can always benefit from a shrewder public relations strategy.

On occasion, Anonymous relied on the classic hoax–lobbing out a lie that in due time would be revealed as a fib to get to a higher truth. For instance, LulzSec hacked and defaced PBS in retaliation for its *Frontline* film on WikiLeaks, *WikiSecrets,* which drew the ire of LulzSec members who condemned the film for how it sensationalized and psychoanalyzed the "dark" inner life of Chelsea Manning, skirting the pressing political issues raised by Wikileaks' release of diplomatic cables. Gaining access to the web server, the hackers implanted fake news about the whereabouts of two celebrity rappers. Featuring a boyish headshot of Tupac Shakur, head slightly cocked, sporting a backwards cap and welcoming smile, the title announced the scoop: "Tupac still alive in New Zealand." It continued: "Prominent rapper Tupac has been found alive and well in a small resort in New Zealand, locals report. The small town–unnamed due to security risks–allegedly housed Tupac and Biggie Smalls (another rapper) for several years. One local, David File, recently passed away, leaving evidence and reports of Tupac's visit in a diary, which he requested be shipped to his family in the United States." Although at first glance it may be unclear why, the defacement delivered a particularly potent political statement. While the fake article and hack caused quite a sensation in the global press, most journalists failed to address LulzSec's criticism of the film's shallow puffery. And yet LulzSec managed to force sensationalist coverage via its hack-hoax combo, instantiating through this back door their original critique of journalists' tendencies to sensationalize news stories.

But in most cases, hoaxing was used sparingly and Anonymous simply amplified messages already being broadcast by other activists or journalists. For instance, one of their most famous operations, #OpSteubenville, concerned a horrific case of sexual assault by members of the high school football team in the small steel-factory town of Steubenville, Ohio. After the *New York Times* wrote an exposé detailing the case, Anonymous continued to hyperactively showcase developments around the Steubenville assault through videos and on Twitter, ensuring its visibility for months until two teenagers were found guilty of rape in March 2013.

Anonymous, like Trump, lured in both the public and the media with splashy acts of spectacle. But Anonymous came together not as a point of individual will to seek credit but as the convergence of a multitude of actors contributing to a multitude of existent social movements, collectives, and organizations. Anonymous flickered most intensely between 2011 and 2015, during a tumultuous period of global unrest and discontent, evident in a range of large-scale popular uprisings across the world: the 15-M movement in Spain, the Arab and African Springs, the Occupy encampments, the student movement in Chile, Black Lives Matter, and the Umbrella Movement in Hong Kong. Anonymous contributed to every one of these campaigns. Their deep entanglement with some of these broader social causes has been commemorated by many who worked with or benefited from Anonymous. In 2011, a photo was shared of

Tunisian children sitting in their school's courtyard, donning white paper cutout Guy Fawkes masks, a gesture of gratitude to Anonymous for bringing the message of their plight to the world. More recently, consider the untimely death of Erica Garner, an anti-police brutality activist and the daughter of Eric Garner, a man who died at the hands of a NYPD officer. Not long after her passing, the person fielding her Twitter account paid their respects to Anonymous: "Shout out to Anonymous... One of the first groups of people that held Erica down from jump street. She loved y'all for real #opicantbreathe."

The point of juxtaposing Trump's lying with Anonymous's truth-telling is merely to highlight that transparency and anonymity rarely follow a binary moral formula, with the former being good and the latter being bad. There are many con men, especially in the political arena, who speak and lie without a literal mask–Donald Trump, Silvio Berlusconi, George W. Bush, Tony Blair–and are never properly held accountable, or it requires a David and Goliath-like effort to eliminate them from power. Indeed, Trump, acting out in the open, is perceived to be "transparent" because he is an individual who doesn't hide behind a mask and, for some, an honest politician for having the bravado to say anything, no matter how offensive. (For some, the more offensive the better.) As sociologist Erving Goffman suggested long ago, humans–so adept at the art of deception–deploy cunning language and at times conniving performance, rather than hiding, for effective misleading.

LESSON 2: THE SHIELD OF ANONYMITY

Transparency can be achieved through existing institutional frameworks, whether by accessing public records, such as using the Freedom of Information Act, or by using the watchdog function of the Fourth Estate. But when these methods fail, anonymous whistleblowing can be an effective mechanism for getting the truth out. Support for this position is cogently articulated in the 1995 Supreme Court case *McIntyre v. Ohio Elections Commission*, which argues that anonymity safeguards the voter, the truth-teller, and even the unpopular opinionator from government retribution or the angry masses of the body

% of Voters in April 2018 Who Believed that Trump Tells the Truth All or Most of the Time

REPUBLICANS ▶

76%

INDEPENDENTS ▶

22%

DEMOCRATS ▶

5%

4,229

◀ Number of false or misleading claims Trump made during his first 558 days in office. At least 122 of those claims have been repeated three times or more.

Average number of false or misleading claims per day. ▶

7.6

Source: NBC News/SurveyMonkey; The Washington Post

politic. The judges of said case wrote, "Anonymity is a shield from the tyranny of the majority.... It thus exemplifies the purpose behind the Bill of Rights and of the First Amendment in particular: to protect unpopular individuals from retaliation... at the hand of an intolerant society." To signal their awareness of and contribution to this tradition, Anonymous participants are fond of quoting Oscar Wilde: "Man is least himself when he talks in his own person. Give him a mask, and he will tell you the truth."

One of the most striking and effective examples that bears out the Supreme Court's rationale and Oscar Wilde's aphorism involves a face mask donned by a medical doctor. In 1972, a psychiatrist presenting at an American Psychiatric Association meeting concealed himself with a voice distorter, pseudonymous name, and rubber mask. Going by Dr. H. Anonymous, and serving on a panel called "Psychiatry: Friend or Foe to Homosexuals?" the doctor opened by confessing: "I am a homosexual. I am a psychiatrist." At the time, homosexuality had been classified by psychiatry as an illness, making it particularly impervious to critique. This bold and gutsy revelation accomplished what Dr. H. Anonymous and his allies had set out to do: re-embolden ongoing efforts to de-pathologize homosexuality. Only a year later, the APA removed homosexuality from its diagnostic manual and Dr. H. Anonymous, who had feared he would not receive academic tenure if his employer found out he was gay, remained protected (and employed), only making his name public twenty-two years later as John E. Fryer.

Many other individuals and groups have spoken and acted truthfully undercover in an attempt to expose some abuse or crime and used anonymity to shield themselves not only from peers, colleagues, or employers, as Dr. Fryer did, but from government retribution. Anonymous, Antifa, Chelsea Manning (during her short tenure as an anonymous leaker), Deep Throat (the anonymous source in the Watergate scandal), and the Citizens' Commission to Investigate the FBI—all of whom have commanded some measure of respect from their words and actions alone, not their legal identities—have delivered transparency that was deemed valuable regardless of their perceived unaccountability or opacity. In the exposure of egregious government wrongdoing, anonymity has the potential to make the risky act of whistleblowing a bit safer. Such was the case with the Citizens' Commission to Investigate the FBI, a group of eight anti-war crusaders who broke into an FBI field office in 1971 and left with crates of files containing proof of COINTELPRO, a covert surveillance and disinformation program levied against dozens of activist movements. The program was eventually shut down after being deemed illegal by the United States government and the intruders were never apprehended. Had these citizens been caught–the FBI dedicated two hundred agents to the case but, failing to find even one of the intruders, gave up in 1976–their fate would have most likely included a costly legal battle followed by time behind bars.

Tragically, people who have spoken unveiled have, at times, been exposed to grave harm and mudslinging. Being honest and transparent, especially when you lack supporters and believers, puts you at risk of a traumatic loss of privacy and, as in the case of Chelsea Manning, physical safety. After being outed by a hacker, Manning was tortured for one year in solitary confinement for her whistleblowing. Former American gymnast Rachael Denhollander, one of the first who dared to call out Larry Nassar, the medical doctor for the U.S. Olympic gymnastics team who sexually assaulted over 260 young women, explained in an op-ed that her life and reputation were ruined for speaking out until the tide began to shift: "I lost my church. I lost my closest friends as a result of advocating for survivors who had been victimized by similar institutional failures in my own community. I lost every shred of privacy." All these examples call to mind the adage "privacy for the weak, transparency for the powerful." Anonymity can fulfill a prescription for transparency by protecting truth-tellers from retaliation.

LESSON 3: EGO CONTAINMENT AND
THE HARMS OF UNBRIDLED CELEBRITY

The rejection by Anonymous of cults of personality and celebrity-seeking is the least understood driver for anonymity, yet one of the most vital to understand. The workings of anonymity under this register function less as a truth-telling device and more as a method for social leveling. Unless you followed

Anonymous closely, this ethos was harder to glean, as it was largely visible only in the backchannels of their social interactions—in private or semi-private chat rooms with occasional bursts on Twitter, such as this tweet by @FemAnonFatal:

> • FemAnonFatal is a Collective • NOT an individual movement NOT a place for self-promotion NOT a place for HATE BUT a place for SISTERHOOD It Is A place to Nurture Revolution Read Our Manifesto... • You Should Have Expected Us • #FemAnon-Fatal #OpFemaleSec

Of course, it's much easier to utter such lofty pronouncements about solidarity than it is to actually implement them. But Anonymous enforced this standard by punishing those who stepped out into the limelight seeking fame and credit. In my many years of observing them, I've witnessed the direct consequences for those who violated this norm. If a novice participant was seen as pining for too much praise from peers, he might be softly warned and chided. For those that dared to append their legal name to some action or creation, the payback was fiercer. At minimum, the transgressor was usually ridiculed or lambasted, with a few individuals ritually "killed off" by being banned from a chat room or network.

Along with punctuated moments of disciplinary action, this norm tended to mostly hum along quietly in the background, but no less powerfully—mandating that everything created under the aegis of Anonymous be attributed to the collective. It's worth stating that, in contrast to their better-known outlaw-hacker compatriots, most Anonymous participants were maneuvering in unambiguously legal territory; those who conjured up compelling messages of hope, dissent, or protest through media like video, snappy manifestos, images, or other clever calls to arms engineered to go viral were not incentivized to anonymity by legal punishment. Moreover, the ethical decree to sublimate personal identity had teeth: participants generally refrained from signing their legal names to these works, some of which surged into prominence, receiving hundreds of thousands of views on YouTube. While a newcomer may have submitted to this decree out of fear of punishment, most participants came to embrace this ethos as a strategy necessary to the broader goals of minimizing human hierarchy and maximizing human equality.

Observing this leashing of the ego was eye-opening. The sheer difficulty of living out this credo revealed itself in practice. As an anthropologist, my methodological duty mandates some degree of direct participation. Most of my labor with Anonymous consisted of journalistic translation work, but on a few occasions I joined small groups of media-makers to craft punchy messages for videos designed to rouse people to action. As an academic writer estranged from the need for pithiness, I recall glowing with pride at the compact wording I once cobbled together to channel the collective rage about some gross political injustice or another. Resisting even a smidgen of credit for the feat was difficult at the time, but in the long run it was satisfying, providing grounds on which to do it again. Still, it not only went against what I've been taught by society, but also the mode of being an academic—someone whose livelihood depends entirely on a well-entrenched, centuries-old system that allots respect based on individual recognition. As the self-named author of this piece, I'd be a hypocrite to advocate a full moratorium on personal attribution. But when a moral economy based on the drive for individual recognition expands to such an extent that it crowds out other possibilities, we can neglect, to our collective peril, other essential ways of being and being with others.

One of the many dangers of unchecked individualism or celebrity is the ease with which it transforms into full-blown narcissism, a personality trait that most obviously forecloses mutual aid, as it practically guarantees some level of interpersonal chaos, if not outright carnage in the form of vitriol, bullying, intimidation, and pathological lying. Trump, again, can serve as a handy reference, as he comes to stand for an almost platonic ideal of narcissism in action. His presidency has demonstrated that an unapologetic solipsism can act as a sort of distortion lens, preventing the normal workings of transparency, truth, shaming, and accountability by offering an aloofness so complete that it seems almost incapable of contemplating the plight of others or admitting a wrong. And in Trump's ascendancy lies a far more disturbing and general lesson to contemplate: that landing one of the most powerful political positions in one of the most powerful nations in the world is possible only because such celebrity-seeking behaviors are rewarded in many aspects of our society. Many dominant cultural ideals enjoin us to seek acknowledgment—whether for our deeds, words, or images. Although celebrity as an ideal is by no means new, there are endless and proliferating avenues at our disposal on the internet to realize, numerically register (in likes and retweets), and thus consolidate and further normalize fame as a condition of everyday living.

To be sure, narcissism and celebrity are far from unchecked. For instance, Trump's conceited, self-aggrandizing traits are subject today to savage critique and analysis by a coterie of pundits, journalists, and other commentators. Even if celebrity is a durable, persistent, and ever-expanding cultural ideal, humility is also valorized. This is true in religious life most obviously, but a bevy of mundane, everyday ethical proscriptions also seek to curb the human ego's appetite for glory and gratification. Something as minor as the acknowledgments section of a book works—even if ever so slightly—to rein in the egoistic notion that individuals are entirely responsible for the laudable creations, discoveries, or works of art attributed to them. After all, it's an extended confession and moment of gratitude to acknowledge that such writing would be impossible, or much worse, if not for the aid of a community of peers, friends, and family. But tales that celebrate solidarity, equality, mutual aid, and humility are rarer. And scarcer still are social mandates where individuals are called upon to hone the art of self-effacement. Anonymous is likely one of the largest laboratories, open to many, to carry out a collective experiment in curtailing the desire for individual credit, encouraging ways to connect with our peers through commitments to indivisibility.

While anonymity can incentivize all sorts of actions and behaviors, in Anonymous's case it meant many of the participants were there for reasons of principle. Their principled quest to right the wrongs inflicted on people embodies the spirit of altruism. Their demand for humility helped to discourage, even if it did not fully eliminate, those participants who simply sought personal glory by joining the group's ranks. Volunteers, compelled into crediting Anonymous, also kept in check a problem plaguing all kinds of social movements: the self-nomination of a rock star or leader, propelled into stardom by the media, whose reputational successes and failures can often unfairly serve as proxy for the rise and fall of the movement writ large. If such self-promotion becomes flagrant, strife and infighting typically afflict social dynamics, which in turn weakens the group's power to effectively organize. The already limited energy is diverted away from campaigns and instead wasted on managing power-hungry individuals.

It's dangerous to romanticize anonymity as virtuous in and of itself. Anonymity online combined with bad-faith actors–pathological abusers, criminals, and collective hordes of trolls–enables behavior with awful, sometimes truly terrifying consequences. Anonymity can aid and abet cruelty even as it can engender nobler moral and political ends–it depends on context. Taking stock of Anonymous's fuller history illustrates this duality. Prior to 2008, the name Anonymous had been used almost exclusively for the purpose of internet trolling–a practice that often amounts to targeting people and organizations for harassment, desecrating reputations, and revealing humiliating or personal information. Having myself been a target in 2010 of a (thankfully unsuccessful) trolling attack, I was thrilled–even if quite surprised–at the dramatic conversion process Anonymous underwent between 2008 and 2010 as they began to troll the powerful, eventually combining the practice with more traditional vocabularies and repertoires for protest and dissent.

As they parted ways with pure trolls, what remained the same was a commitment to anonymity, used for different ends under different circumstances. Still, a number of Anonymous's operations serving the public interest, such as the wholesale dumping of emails that breached people's privacy, were carried out imperfectly and are worthy of condemnation. These imperfect operations should not nullify the positive aspects that the group achieved through anonymity, but should nevertheless be criticized for their privacy violations and used as examples for improving their methods.

Preventing the state from stamping out anonymity requires strong rationales for its essential role in safeguarding democracy. In defending anonymity, it is difficult to simply argue, much less prove, that the good it enables outweighs its harms, as the social outcomes of anonymity are hard to tally. Notwithstanding the difficulties in measurement, history has shown that nation-states with unchecked surveillance power drift toward despotism and totalitarianism. Citizens under watch, or simply under the threat of surveillance, live in fear of retribution and are discouraged from individually speaking out, organizing, and breaking the law in ways that keep states and corporations accountable.

Unequivocally defending anonymity in such a way doesn't make all uses of anonymity by citizens acceptable. When assessing the social life of anonymity, one must also ask a series of questions: What is the anonymous action? What people, causes, or social movements are being aided? Is it punching up or down? All of these factors clarify the stakes and the consequences of using the shield of anonymity. It invites solutions for mitigating some of its harms instead of demanding anonymity's elimination entirely. Technologists can redesign digital platforms to prevent abuse, for example by enabling the reporting of offending accounts. Recognizing anonymity's misuse is why we also ensure limited law enforcement capacity to de-anonymize those who are using cover for activities society has deemed unconscionable, like child pornography. As it stands now, the state commands vast resources, in the form of money, technology, and legitimacy, for effective law enforcement. To additionally call for ending strong encryption, adding back doors for government access, or banning anonymity

The Intercept   intrcept32ncblef.onion

The New York Times   nyttips4bmquxfzw.onion

The Guardian   33y6fjyhs3phzfiji.onion

The New Yorker   icpozbs6r6yrwt67.onion

△ DATA   Tor Addresses for Sending Anonymous Tips

tools—something the FBI often does—is to call for the unacceptable elimination of the many legitimate uses of anonymity.

In spite of these justifications, it is difficult to defend anonymity when some people have only an inchoate sense of anonymity's connection to democratic processes, or see no need for anonymity at all, and others see it only as a magnet for depraved forms of criminality, cowardice, and cruelty. I was reminded of this very point recently after running into one of my former students while traveling. Surprised to recognize me in the group with whom she was about to go scuba diving, she gleefully identified me by subject of study: "You're the hacker professor!" A few hours later, as we climbed out of a small skiff, she asked me unprompted to remind her of my arguments against the common dismissal of privacy and anonymity on the grounds of the speaker "having nothing to hide." I chuckled, given that my mind was occupied with these very questions as I was puzzling through this article, and rattled off a number of the arguments explored here. I'm unsure whether the precise arguments escaped her because years had elapsed, because my lecture was boring, or because the merits of anonymity are counterintuitive to many; likely it was some combination of all three. Regardless, I was pleased that she even had the question on her mind.

It was a reminder that, at a time when examples of anonymous actors working for good aren't readily available in the news, as they were during the days of Anonymous, those of us attempting to salvage anonymity's reputation need to put forward compelling tales of moral good enabled by anonymity, rather than exploring it only as some abstract concept, righteous on its own, independent of context. Anonymous remains an exemplary case study to that aim. Aside from using the shield for direct action and dissent, for seeking truth and transparency, Anonymous has also provided a zone where the recalibration of credit and attribution has been not just discussed but truly enacted. In doing so, Anonymous provided asylum from the need to incessantly vie for personal attention, becoming notorious while tempering individual celebrity, and yet still managed to fight injustice with spectacle, all while standing anonymously as one. ●

It Takes a Village

Camille Fassett

n July 2017, residents of a Boston neighborhood spotted police officers flying a drone around a housing project. Jamaica Plain resident My'Kel McMillen saw it for himself. He took photographs that clearly showed the drone, Boston Police Department squad cars, and officers in uniform, and sent them to his local ACLU chapter.

Kade Crockford, Director for the Technology for Liberty Program at the ACLU of Massachusetts, was shocked. Days later, Crockford filed a public records request with the BPD to find out why officers were flying a surveillance drone in a residential, predominantly black area.

The department responded that, while it had purchased three drones, none had been flown that day. It claimed the drones had never even been unboxed. Despite numerous photographs, witnesses, and news articles about the incident, the department repeated its assertion that the drones had never been used.

"There was no information communicated that there would be drones in my community, and no notice from the Boston Police Department that they were testing them," McMillen said. "No one knew what they were filming or recording."

In 2015, the BPD fought public records requests by two individuals for information related to the department's use of cell phone surveillance technologies before finally confirming it did possess such a system. Before Crockford's request, the BPD had not disclosed that it had drones, or that it was even considering purchasing them. It had no legal requirement to do so.

Like the BPD, other law enforcement agencies are rapidly acquiring new surveillance technologies, and are in many cases not required to gain approval from local government or even to inform the public.

When the police department in San Jose, Calif., purchased a drone, it did not inform the public either prior to or after the purchase. It used nearly seven thousand dollars in federal grant money to purchase it—money intended to aid the department's bomb squad in assessing threats like explosives. Oftentimes departments avoid scrutiny and undermine local democratic processes by obtaining new policing equipment through federal funds. The Department of

Homeland Security, for example, showers millions of dollars in grant money on local law enforcement agencies.

Drones, or unmanned aerial systems, can carry heat sensors and cameras, including infrared. Some can track people and vehicles at altitudes of twenty thousand feet. Police use them for routine law enforcement purposes, such as locating missing persons and assessing damage to homes, bridges, power lines, and oil and gas facilities after natural disasters. But some have also proposed using them to record traffic violations and to monitor protests. One bill in Illinois, which thankfully has been stopped for now, would have allowed police to deploy drones armed with facial recognition software at any gathering of more than one hundred people.

Law enforcement agencies are required to obtain a warrant to use a drone in only approximately one third of states. And because of their vast technological capabilities, such as flying far above the limits of human sight, drones could be used to spy on Americans without their knowledge, even if a warrant is required.

But the morphing landscape of high-tech electronic surveillance goes far beyond drones. Perhaps the most common spying equipment used by police departments, surveillance cameras can allow police to monitor people virtually everywhere. A number of police departments are also allowing private businesses and residences to grant them access to their camera footage, expanding police surveillance into private spaces.

Some agencies have purchased automated license plate readers (ALPRs), which capture every license plate number that they pass and link them to information like the location, date, and time they were logged. Although the data collected by ALPRs is not attached to a person's name, this information can reveal identifying driving patterns, what vehicles visited particular locations, and a driver's associates. Much of this data is consolidated in a database that can be accessed by agencies across the country.

Facial recognition software could be used to identify people from pictures or photographs, in real time or even at traffic stops. And systems of audio sensors, often sold under the brand name ShotSpotter, alert police to the

Drone Law in
the United States

Wisconsin

North
Dakota

Iowa

Illinois

Maine

Oregon

Idaho

Montana

Vermont

Indiana

North
Carolina

South
Carolina

Nevada

Utah

Florida

Texas

Tennessee

Source: National Conference of State Legislatures

States that do not require warrants
for any type of drone activity.

Arkansas, California, Kansas,
and Michigan require warrants for
non-government use only.

sound of gunfire. In order to do so, the systems necessarily listen at all times.

Cell site simulators–commonly known as Stingrays–are particularly controversial surveillance devices that mimic cell phone towers so that police can identify and track mobile devices nearby without ever involving phone companies. Some can gather information including unique data about a given device and even the content of unencrypted calls or messages.

These technologies, particularly combined, can exponentially increase the capacity of police departments to watch their citizens. For example, drones can carry other kinds of surveillance technologies–like cell site simulators and cameras with facial recognition software–amplifying the capabilities of both.

These vast powers, largely devoid of oversight, disproportionately burden already disadvantaged people. In 2012, the New York Police Department captured the license plate numbers of every car parked near a mosque. U.S. Immigrations and Customs Enforcement (ICE) has access to a national database that allows the agency to track immigrants. Indiscriminate mass surveillance specifically impacts people who visit sensitive locations, like women who visit reproductive health clinics, and those who are already targeted by police–largely Muslim, black, and brown people.

But across the United States, communities concerned about secret surveillance are fighting to change that by enacting local legislation that would require transparency and accountability in law enforcement acquisition and use of spying technologies.

Santa Clara County in Silicon Valley and Seattle made history by becoming the first jurisdictions in the United States to enact surveillance ordinances, in 2016 and 2017 respectively. In both regions, agencies are now required to involve the public, gain approval from officials before acquiring new surveillance technologies, and develop use policies that include civil liberties protections.

Santa Clara County Supervisor Joe Simitian proposed the Santa Clara legislation, which mandates annual surveillance reports that detail how the technologies were deployed throughout the year. It also requires that existing technologies that were acquired before the ordinance was adopted have use policies put in place.

For example, the January 2018 surveillance report on the sheriff office's use of body-worn cameras (BWCs) goes into detail about the costs and data sharing of the technology. It notes that "there were 861 files of evidence" shared with the District Attorney's office that year, and "four California Public Records Act requests for BWC footage." Once approved, the surveillance reports are available online for any member of the public to view.

"The ordinance does not prohibit the acquisition or use of any technology, but rather it requires thoughtful and thorough discussion, and policies are approved by the board of supervisors before acquiring it," Simitian said.

Just like in Boston, in 2010 the City of Seattle had acquired drones with no public process and no articulation of any policies governing usage and data retention. The strong opposition that resulted when the purchase of the two drones became public not only led to a halt to the drone program, but also sparked efforts towards creating what would become one of the country's first surveillance ordinances.

Shankar Narayan, Technology and Liberty Project Director for the ACLU of Washington, said that the first version of the ordinance that was passed in Seattle later that year was ineffective, left loopholes, and lacked enforcement mechanisms. He believes that the second version of the ordinance, which was passed in 2017 despite pushback from police, contains stronger language that remedies these problems and clearly defines surveillance technology.

The legislation states that it intends to "govern the acquisition and use of surveillance equipment" and that the city's use of surveillance technologies "should include specific steps to mitigate civil liberties concerns and the risks of information sharing with entities such as the federal government; and should incorporate racial equity principles into such protocols."

Future attempts by Seattle agencies to acquire any new technology (except cameras) that would be primarily used to observe or monitor citizens "in a manner that is reasonably likely to raise concerns about civil liberties, freedom of speech or association, racial equity or social justice" will undergo a rigorous approval process that includes community input. City Council review of all surveillance technology possessed by Seattle began in March of 2018.

Santa Clara County's and Seattle's ordinances are powerful examples of how jurisdictions nationwide can proactively address mass surveillance and maintain transparency. Since their implementation, several regions are considering similar legislation.

These efforts are part of the ACLU's Community Control Over Police Surveillance (CCOPS) campaign. Launched in September 2016, it is a partnership between the ACLU and other national organizations, activists, and community members, many having fought for a seat at the table in police decision-making processes for years.

The CCOPS campaign has created a model bill for cities interested in passing legislation that applies transparency and public empowerment principles to government agency acquisition processes. The model bill was developed by a coalition of civil liberties groups around 2014, and has been enhanced over the years in response to feedback from different cities that have adopted it, according to Tracy Rosenberg, a member of activist group Oakland Privacy and Executive Director of Media Alliance.

"The City Council finds it is essential to have an informed public debate as early as possible about decisions related to the funding, acquisition, and deployment of military and surveillance equipment by local law enforcement," begins the ACLU's CCOPS+M model bill, which also covers the use of military equipment.

In a post to its national blog about CCOPS, the ACLU wrote that "the effort's principal objective is to pass CCOPS laws that ensure residents, through local city councils, are empowered to decide if and how surveillance technologies are used, through a process that maximizes the public's influence over those decisions." It continues, "Passing CCOPS laws would empower city councils to say 'no' to secret surveillance sharing agreements between the feds and local police."

Around the time of the Snowden revelations in 2013, the City of Oakland, Calif., had been working toward constructing a massive surveillance project. It would have comprised a network of over seven hundred cameras, facial recognition software, license plate readers, and a sensor that identifies gunshots

The website freedom-to-tinker. com, hosted by Princeton's Center for Information Technology Policy, published a study highlighting a particularly invasive data-mining software called "session replay scripts" that are being used by an increasing number of websites. According to the study, session replay scripts "record your keystrokes, mouse movements, and scrolling behavior, along with the entire contents of the pages you visit." Unlike most third-party analytics services, which provide aggregate statistics of your searches and the pages you visit, session replay scripts actually record your individual browsing session in its entirety, "as if someone is looking over your shoulder."

The study lists tens of thousands of websites that were either found recording users' browsing sessions or have the capability to do so. Among the big-name sites are xfinity.com, windows.com, texas.gov, petco. com, and fandango.com. The following sites were also found on that list.   —Landon Bates

openme.com
everyinteraction.com

realclearlife.com
rainbowlight.com

springhappenings.com
hometocome.com
wallsneedlove.com

mostlypaws.com
bunnycart.com
borrowmydoggy.com

nutsaboutgranola.com
aloyoga.com
guavajuicebox.com

drinkmorewater.com
saturatetheworld.com

kissflows.com
fluidmeet.com
beget.com
itmydream.com

babylegs.com
daydreamdaycare.com

mykidneedsthat.com
bringmethat.com
pissedconsumer.com
fondofbags.com
wealthify.com
salesoptimize.com
opulentitems.com

statuscake.com
globegain.com
titswanted.com
sexlikereal.com
oopscum.com
cumexposed.com
cumrainbow.com

hotrawsex.com
sexyrealsexdoll.com
myfirstdrone.com
oldmaturepussy.com

sweetpiss.com
germangoogirls.com
disgustingmen.com
mom-fuck.me

guardyoureyes.com
hidemyass.com

dontfear.ru
seekurity.com
webtegrity.com
securethoughts.com

neurohacker.com
thatspersonal.com

thinkingphones.com
lossofsoul.com

publicdesire.com
transparentcontainer.com

wearecontent.com

by sound. A group of concerned citizens quickly mobilized against this effort and successfully limited Oakland's adoption of these technologies. The city council eliminated the project's particularly concerning elements, like facial recognition software and data retention, and created a landmark committee of citizens to oversee the remaining pieces of the project.

This citizen committee became a privacy commission that advises the city council, and years later has championed an ordinance mandating transparency and public involvement in law enforcement use of surveillance equipment. In addition to this new approval process, the ordinance also includes whistleblower protections.

"The whistleblower protections affirmatively state retaliatory actions, including personnel actions, are prohibited against employees or job applicants for carrying out the ordinance's provisions or complaining about the failure to do so. They provide specific injunctive relief and, when warranted, financial damages," said Rosenberg.

Brian Hofer, a member of the Oakland Privacy Commission, sees Oakland's privacy ordinance as a model that any city could replicate and modify so that each community can determine what levels of government intrusion it is willing to allow. "The beauty of this ordinance is that any city could use the same framework, and come to different conclusions in increasing transparency, and, hopefully, prohibiting misconduct," he said.

On March 13, 2018, the Berkeley City Council unanimously approved its own privacy ordinance, setting a strong precedent for surrounding cities. The City of Davis, Calif., followed on March 20. Other jurisdictions in the San Francisco Bay Area—including Alameda County and Bay Area Rapid Transit—are considering adopting similar ordinances based on the CCOPS model bill. Beyond California, Nashville, Tenn., and Somerville, Mass., have adopted ordinances like Oakland's, and similar efforts are underway in Cambridge, Mass., and St. Louis, Mo.

So far, proactive technology oversight measures have mostly been undertaken at the local level. But in California, a bill is underway that would enact surveillance technology reform at the state level: Senate Bill 1186, introduced by California Senator Jerry Hill.

SB 1186 aims to ensure that community members have a seat at the table in decision-making processes regarding police use of surveillance technology. It requires public debate and a vote by local elected leaders prior to law enforcement's acquisition of new surveillance technology. Matt Cagle, Technology and Civil Liberties Attorney at the ACLU of Northern California, said that "SB 1186 makes sure the right questions are asked and answered about surveillance technology from the beginning and smart decisions are made that keep communities safe."

Cagle said that state legislation like SB 1186 "reinforces and builds on the progress that cities like Oakland and Davis have made."

SB 1186 would require each agency to "submit to its governing body at a regularly scheduled hearing, open to the public, a proposed Surveillance Use Policy for the use of each type of surveillance technology and the information collected, as specified," and "prohibit a law enforcement agency from selling, sharing, or transferring information gathered by surveillance technology, except to another law enforcement agency, as permitted by law and the terms of the Surveillance Use Policy." SB 1186 also provides that people can sue agencies that violate such policies and recover attorneys' fees.

Similar legislation—Senate Bill 21—died last session after clearing the state senate. But Cagle thinks that this movement is gaining steam, and is confident that this time around the bill will get to the governor's desk. He noted that SB 1186 is supported by a "broad statewide civil rights coalition including organizations fighting for racial justice and immigrants' rights."

"It would be great to see something like California's Senate Bill 21 in Washington as well," Narayan said, emphasizing that there is a role for legislation promoting fairness and community input in technology adoption at local, state, and federal levels.

"There should be a local process for communities to articulate concerns that hit close to home," Narayan told me. "But there are things that can be regulated by the federal government, like broadband privacy. Ideally, we see laws that interact with each other."

Jurisdictions like Santa Clara County and Seattle aim to enact ordinances that are flexible and broad enough to address a wide diversity of surveillance equipment, from cell site simulators to facial recognition software, and the rapidly evolving nature of technology. In other jurisdictions, communities are fighting for more specific legislation that responds to specific technologies being deployed.

In Northampton, Mass., the police department proposed drastically increasing the number of surveillance cameras in the city's downtown area. Concerned by this expansion of police power, privacy advocates drafted an ordinance that established restrictions on the use of cameras in public areas. The police chief had admitted that the department would turn over images captured by their cameras to any law enforcement agency that requested them, including ICE. "This would undermine the sanctuary city designation that we've embraced by allowing ICE to use images upon request," said William Newman, an attorney with the ACLU of Massachusetts.

Newman noted that the threat of expansion of cameras sparked awareness among the public of the dangers of surveillance. "This debate has opened up the potential for Northampton to more widely address surveillance technology," he said.

After the city council passed an ordinance limiting the use of city-owned cameras in Northampton's business district, the mayor vetoed it. In a strong defense of the privacy of Northampton residents, the city council overrode the mayor's veto and voted to uphold the ordinance in January.

Crockford agrees that people care more about their right to privacy than ever before, and cites Edward Snowden's 2013 revelations as a significant reason people are aware of government surveillance. "Privacy is a negative value–many people only realize it's valuable when taken away, like a hidden necessity."

And as the fight for privacy rights gains momentum, so, too, does the movement for privacy ordinances.

In contrast to strong policies mandating community involvement and police accountability, the current status quo puts the burden on the public to fight with government agencies to glean information from public records requests and whistleblowers.

McMillen says that without use policies and communication between the city and residents, the community can't hold the government accountable for its use of drones or any other type of surveillance technology. "It's intimidating, especially to members of the community who are afraid to speak up for fear of law enforcement retaliation."

Santa Clara's ordinance requiring transparency in surveillance technologies has been enforced since 2016. Since then, some equipment proposals have been adopted, some rejected, and some modified after feedback or criticism. The Santa Clara County Sheriff's Department submitted a use policy and a surveillance-impact statement as part of its proposal for officer BWCs. After civil liberties groups initially opposed the broad draft language, the Department updated the text to address some of their concerns. Ultimately, the proposal was approved.

But after hosting public debate on a different proposal, as required by the ordinance, the Santa Clara County Board of Supervisors placed a moratorium on facial recognition use with police officer BWCs.

"[The ordinance] has forced people to answer questions like, When and how will we use this? If we acquire data, who will keep it, and who will have access? These questions weren't asked, let alone answered, before it was required," Santa Clara County Supervisor Simitian said.

Privacy ordinances do not necessarily stop police departments from obtaining or misusing surveillance technology. But legislation is one possible tactic or methodology of pushing back against mass state monitoring, and pushing for greater transparency. "The solution is changing the way we define public safety, and a number of associated concepts like the nature of policing, the definition of terrorism, and the assumption that collecting the maximum amount of data technologically possible on people, regardless of any wrongdoing, is an unquestionable good," said Rosenberg.

If a surveillance ordinance like Santa Clara's or Seattle's had been in place in Boston, the BPD would have had to gain approval from an elected body before acquiring drones, let alone flying them in Jamaica Plain. The community would have had a chance to fight against their use before they were obtained, and would have measures in place to hold law enforcement accountable if and when the tools were misused.

"Never once has a police state failed to use technology and surveillance to control a population. Never once has a ruling party conducted mass surveillance without leading to great harms," Hofer told me. "Unfettered use always leads to bad things, and not one case breaks that–there is no friendly test case for mass surveillance. It always leads to humanitarian and civil liberties abuses." ●

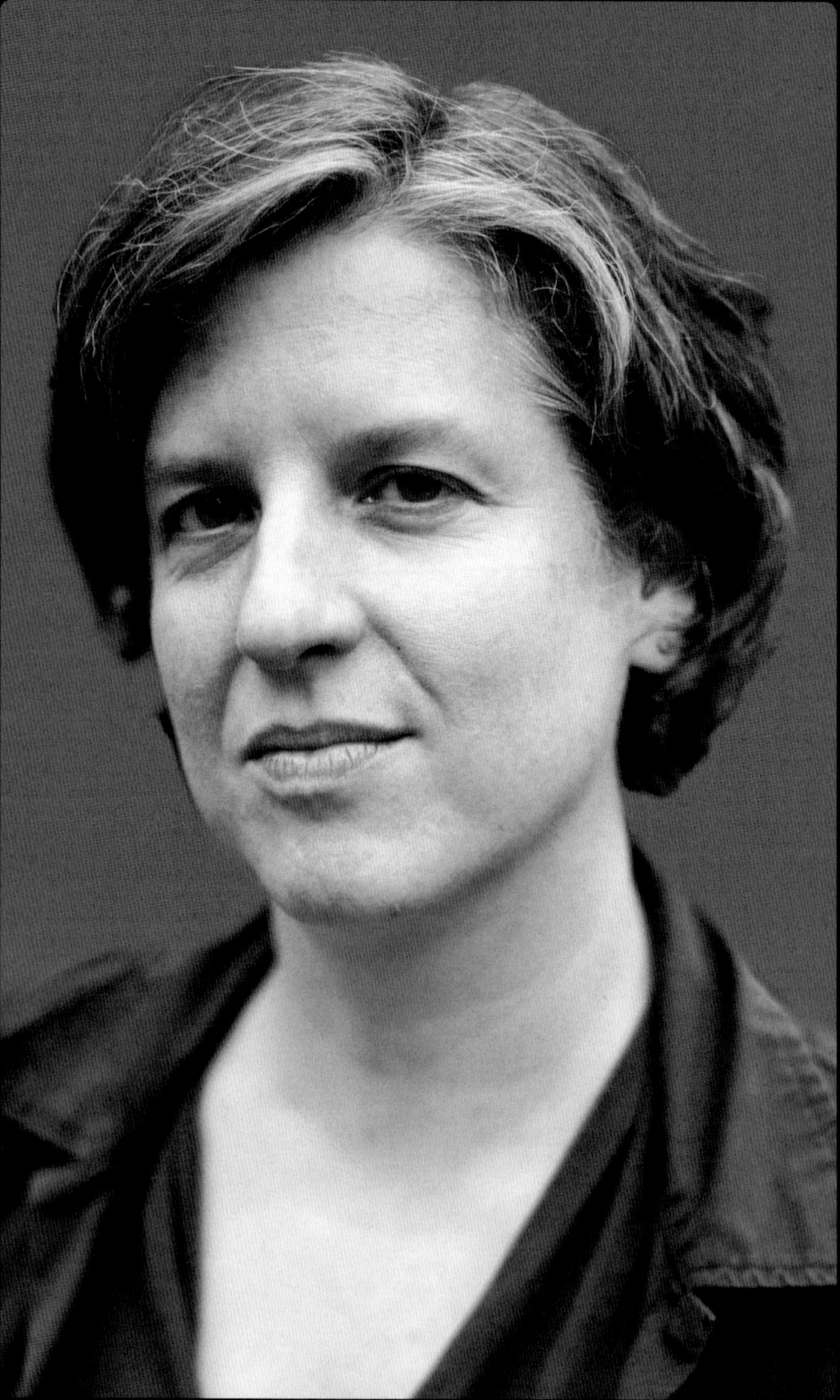

Photo: Sadaf Rassoul Cameron

"Once you're in the system, you're in forever."

Virginia Eubanks talks with Jacob Silverman

irginia Eubanks's new book, Automating Inequality: How High-Tech Tools Profile, Police, and Punish the Poor, *is a study of the injustices and absurdities that come with digitizing various social-service programs. The subject might seem academic, even picayune, but the book's achievement is to show that we all, sadly, live in what Eubanks calls "the digital poorhouse."* As Eubanks shows, the choices we make as a society about how we treat the poor have wide, culture-defining effects. (Eubanks learned this firsthand when her partner's medical emergency led her down a money-draining bureaucratic spiral that included being accused of health-insurance fraud.) The programs she critiques affect nearly all Americans in some way, making them all the worthier of improvement, especially at a time when the right is endeavoring to defund health care and other social programs.

Drawing on her scholarly expertise in women's studies and political science, Eubanks's attentions range from welfare benefits in Indiana to housing assistance in Los Angeles to child protective services in Pittsburgh. Invariably, these programs privilege computer software over human intervention (and human empathy), criminalize poverty, and make benefits more difficult to obtain. There are also instances of corporate corruption, incompetence, and sheer blindness to the actual needs of working and marginalized people.

"Technologies of poverty management are not neutral," Eubanks argues. "They are shaped by our nation's fear of economic insecurity and hatred of the poor; they in turn shape the politics and experience of poverty."

In the course of her book, Eubanks shows that these technologies of automation—along with the political assumptions underpinning them—tend to harm the poor more than they help. By establishing a new kind of data-based surveillance, one in which welfare recipients are under constant scrutiny and may be penalized for any contact with the system at all, these systems divest working people of their privacy and autonomy. They exert a punitive hold over people they're supposed to support, and ensure that the price of accessing basic services is the complete surrender of one's personal data.

I recently talked to Eubanks by phone about the digital poorhouse, the

criminalization of poverty, and how we might fight back against these dehuman-
izing systems. Our interview has been slightly condensed and edited for clarity.

JACOB SILVERMAN: You're venturing into very novel territory here. How would you describe your beat?

VIRGINIA EUBANKS: I say that I write about technology and poverty, and one of the things that's interesting about describing it that way is that people tend to immediately assume I'm talking about the lack of access for poor and working people to the newest tools of the technology revolution. For twenty years, my work has pointed out the opposite: that poor working communities tend to be the target for some of the newest—and in the recent work I've done—most invasive and punitive technology innovation.

So, part of my work is helping people understand that the social-justice issues around technology go beyond access, go beyond what in the past would have been called the digital divide. Though access is still really important, there's a whole world of other issues around targeting and policing specifically that we need to keep on the social-justice agenda.

JS: I've seen some comments from you about wanting to go beyond the tra-ditional definition of privacy to talk about autonomy or personal agency. Can you talk about what you mean by that?

VE: The reason I'm trying to push a little bit on these ideas of privacy is that for the folks who I talk to the most—people who are unhoused, folks who are on public assistance, folks who are being investigated or who have some kind of contact with child welfare systems in this country—privacy is just not the first thing that comes up in their list of concerns. It's something that people care about, but it's not really even in the top five. What folks are telling me is that in their lives, this expectation of privacy is very much influenced by the amount of economic or political power that folks have. In the public assistance system, for example, there's just not an expectation of privacy. The historical legacy of public assistance in the United States is that you trade some of your basic human rights, including your right to privacy and your right to tell your story in your own way, for access to public benefits. That's part of the context for people that come into contact with these systems, and I believe that's unjust.

So I actually find, when most people talk about what they want, that they want self-determination. They want the ability to make decisions for themselves about the things that most impact their lives, like how they spend their money, where they live, whether their family stays together. Those are things people want to be self-determining around. Privacy might be part of that, but it's not necessarily an expectation or a priority for folks that I talk to.

JS: Would you still use the term "surveillance" when talking about a welfare system that collects loads of information on people? I noticed you used some form of the term "data-based surveillance" in your book. Would you describe this as surveillance of people who requested these services, or is it something else?

VE: I don't think I use the word *surveillance* a lot. One of the things I hope my work will accomplish is for us to think more expansively about policing. The processes of policing, of putting people into boxes so they're more manageable, are processes that happen across all sorts of different public services, so if you look at child protective services, or homeless services, or public assistance, there're processes of policing happening in all of those areas. So I think one of the reasons that I don't always use the word *surveillance* first is that it has a tendency to create an image in people's minds of a police officer sitting in a car across the street from your house. One, that's not really what surveillance looks like anymore, and two, I want to invite people to look at these processes of policing beyond the police.

I wouldn't say it's *not* surveillance. I just don't know that that's the most evocative term for what's happening.

JS: One thing that stuck out to me from the book is the notion that there *is* a lot of data being collected, and once you're in the system, you're in forever. If something happens to you that draws the attention of social services or law enforcement, they might dig up your past and ask you to explain, for instance, some visit from child protective services that occurred years ago because your angry neighbor decided to call them on you. Is that something different, this infinite and meticulous record keeping? Or is there a longer lineage to this kind of practice?

VE: The point of using the metaphor of the digital poorhouse in the book is that we have a tendency to talk about these new tools as if they just appeared from nowhere–the monolith from *2001: A Space Odyssey* that just arrives from space. The point of using that metaphor is to ground those new tools in history, and even though we talk about them as disruptive, they tend in public services to be more evolution than revolution. The key origin story for these tools, for me, is the establishment of the actual brick-and-mortar poorhouse, which was an institution for incarcerating the poor.

So the poorhouse movement rose in the United States in response to a really huge economic crisis–the depression of 1819–and specifically to poor people's organizing around their rights and their survival in response to that depression. The logic of the poorhouse is really important because it's the deep social programming that underlies a lot of these tools. The logic is basically that there's this distinction between poverty, which is a lack of resources, and pauperism, which is dependence on public benefits. And this should sound familiar: the idea among economic elites is that the problem is not poverty, even in the midst of a massive depression; the problem is dependence.

So, the solution that the poorhouse offered was that, in order to receive help, you had to agree to basically enter into this voluntary prison–this very frightening, very dangerous institution. You had to give up a bunch of your civil rights, if you had them, if you were white and male. Your right to marry, your right to vote. You were often separated from your children. And often you risked your life, because the death rates at some poorhouses was 30 percent annually, so a third of the people who entered died every year. This is a moment in our history where we as a nation made this decision that public services should be more of a moral diagnosis than a universal provision of resources; it's more of a moral barometer, and less of a universal floor. That's the deep programming that we see coming up in these systems over and over again.

And then you ask about the eternal record, specifically whether this is brand new, and in many ways it's not. One of the origin stories of this book is that I went looking for what year New York state welfare systems went digital. I thought that would be in the 1980s or the 1990s, and in fact I ended up going way far back in the New York state archives, all the way back to the register books of the county poorhouses. There's a lot of continuity in these systems, going all the way back to the 1800s. But this practice of collecting a massive amount of data on people who are living in situations where their rights are not necessarily being respected: that's an old idea. The form it's taking now is quite different.

The metaphor of the digital poorhouse only goes so far, because there are things about the poorhouse that are not possible in the digital poorhouse. Like, one of the things that's interesting about the poorhouse is it was an integrated institution: old folks, young folks, people of all races, from all different places, all literally sleeping together in the same rooms, and eating together at the same table. And one of the outcomes of the actual county poorhouse is people actually found alliance and solidarity, whereas the digital poorhouse is really about separating and isolating people from each other, because part of the way it works is by surveilling your networks, which may make people isolate themselves in response. The county brick-and-mortar poorhouse had records but, with rare exceptions, they were paper records that eventually crumbled away to dust. Now, we have what's potentially an eternal record. Which is not to say it's not really hard, as someone who knows something about archiving digital records—let's not overstate how eternal these things are.

JS: You write that we all live in the digital poorhouse. Is that a reflection of the notion that many people go through a crisis of poverty or other precariousness and have to deal with these systems, or is there some other way that we all live under this system?

VE: The first thing is when I say poor and working-class people, it's really important for people to understand that one of the narratives I'm pushing against is the idea that poverty only happens to a tiny minority of probably pathological people. The facts are that in the United States 51 percent of us will dip below the poverty line between the ages of twenty and sixty-four, and a full two-thirds of us will access means-tested public assistance. And that's not social security or free school lunches, that's straight welfare. When I say poor and working people, I'm talking about the majority of people in the United States, who will at some point rely on these systems. So, two-thirds of us will be in these systems in a really direct way.

The second thing I'm trying to point out by saying "we all live in a digital poorhouse" is that we've always all lived in the world we create for poor people. The United States is very much the exception in that we don't follow human rights approaches to public provision of shared social goods. We have created this set of programs that is expensive, punitive, paternalistic, and often criminalizing of people who are suffering economic shock or setback or illness.

When we see the general lack of health and wellness in the United States, we all live in that because we've created these health systems that are so hard to navigate, and they are based on the free market rather than on the basic provision of human rights, like medical care. We've all always lived in this world where we accept a dysfunctional culture because we want to punish poor folks.

And also, frankly, that has a lot to do with race and the history of race in the United States. Part of it is not just that we don't want poor folks broadly to have access to public services in the United States; white Americans don't want people of color to have access to shared public resources. So we've created this really punitive culture around public services, which impacts all of our lives in these really deep ways. The kinds of communities we live in, the sort of schools we have, the kinds of medical care we get—that's all impacted by this punitive attitude we have towards economic inequality.

JS: Something that really struck me was a comment you make in the book about how seeking out many of these services is lauded in one way or another, considered a mark of good self-care, if done in the private sector–services like drug treatment or therapy or after-school programs. But then when done through public means, through welfare or social services, people are often looked down upon in some way. Is this that attitude of seeing poverty in punitive terms?

VE: Yeah, I say in the book, specifically when I'm talking about the Allegheny Family Screening Tool, which is this statistical model that's supposed to predict which children will be victims of abuse and neglect in the future, that these systems often confuse parenting while poor with poor parenting. For many poor and working-class families, because we've so limited access to cash benefits and food stamps and public housing that the child welfare system is often the resource provider of last resort. Because they can also pull some strings. Like, if you're being investigated for child neglect, your caseworker might be able to get you into public housing. Or, they might be able to lean on your landlord to do repairs to your house that, if unattended to, could result in you losing your child for child neglect. It's just this huge, horrifying irony, that this is the system of last resort for many families, but the only way you can get involved in it is by giving the state the power to break up your family. So, by asking for these resources, people feel like they often have to trade their right to family integrity.

One of the women I talked to in Pittsburgh, Allegheny County, said they'd help you but you have to give in first. You have to put your child into the system, she feels, before they help you. And that orientation carries through into the design of these systems, so the Allegheny Family Screening Tool is built on a data warehouse designed in 1999 that now holds a billion records, which is eight hundred records for every individual living in Allegheny County. But it doesn't, of course, collect data on all the individuals in Allegheny County; it only collects information on those folks who are interacting with public services. So, if you ask for mental health or addiction support from a county service, you end up in there. But if you pay for addiction support or mental health support through private insurance, you don't end up in there. If you ask for help with childcare or respite support from a county service, you end up in the data warehouse. If you can afford a nanny or a babysitter or an au pair, you don't.

So that structure builds in the assumption that asking for support from shared public resources is a sign of risky parenting, when in fact it's often a sign that you're doing exactly what you need to do. You're getting the support you need to raise a happy, healthy, and safe family. That's one of my great fears, that these systems will target those folks who are really doing the best they can to provide for their kids and be great parents, and may even result in families not reaching out for those resources they need, becoming increasingly isolated, and lacking services. According to the CDC, lack of resources and social isolation are two of the key predictors for child maltreatment. So it may in fact create the very thing that they most fear.

JS: You made reference to the fact that the systems these services are built on can often be political in some way. You also write in your book that these systems "preempt politics." Can you speak a little bit about how that happens–how do politics get abstracted away?

VE: We tend to believe that our digital decision-making tools are objective and neutral. But one of the big arguments of the book is that there are actually all these moral and political assumptions built right into them, but they're often hard to see because they get talked about as basic administrative changes, not as policy changes. Let me give you a concrete example of that. One of the cases I talk about in my book is this attempt to automate all of the eligibility processes for the public assistance system in the state of Indiana. That's cash welfare, Medicaid, and at the time what was called food stamps, and is now called SNAP. And the system is based on this belief that personal relationships between caseworkers and recipients were invitations to collusion and fraud. So, when the governor went around the state talking about the system and the $1.4 billion contract that the state signed with IBM, at every stop he talked about this very famous case in Indianapolis where a number of caseworkers had colluded with recipients to defraud the government of, I think it was about $8,000. He mentioned this at every whistle stop of this tour.

So, part of this system design was severing that connection between caseworkers and the families that they served, and what that looked like for Indiana was moving fifteen hundred local caseworkers, who in the past had been responsible for a docket of families, to private call centers where they became responsible for, rather than families, a queue of tasks as they drop into their computer system. What that meant was that no one person was responsible for a case from beginning to end, and every time a recipient called one of these call centers, they talked to a different case worker, so they had to start over every time. And the result of this was a million denials of benefits in the first three years of that project, which was a 54 percent increase in denials from the three years before. And most people were denied for this reason the computer called "failure to cooperate in establishing eligibility"– which basically accounted for any mistake that anyone in the system made, whether someone forgot to sign page 34 of this really lengthy application, whether someone in the call center made a mistake, whether someone in the document scanning center made a mistake–all of those were interpreted as active failures to cooperate on the part of the person applying for services.

And that had a huge impact on people's day-to-day lives. Folks who were denied included a six-year-old girl with cerebral palsy, Sophie Stipes, from a white rural town in central Indiana–I write about her and her family in the book–or Omega Young, an African American woman from Evansville, Indiana, who missed a recertification phone call because she was in the hospital suffering from terminal ovarian cancer and was kicked off her Medicaid because of it. The lesson here is really that we talk about these systems as simple administrative upgrades, but often they contain really important political decisions. In Indiana, the political decision was that it is more important to block people who might not be fully participating in the process from getting resources than it is to extend a little bit of extra help to make sure folks who are really really vulnerable are getting the resources that they are entitled to and legally eligible for.

Chris Holly, a Medicaid attorney and one of the people I interviewed in Indiana, put it really beautifully when he said that we have a justice system where supposedly, philosophically at least, it's more important to us that ten guilty people go free than one innocent person serve time in jail. But this system in Indiana flips that on its head. It was more important to the state that one ineligible person not get resources than it was that ten eligible people be denied. And that's a huge political shift–that's a political decision. We often hide these kinds of political decisions behind the cover of these supposedly neutral and objective technologies.

JS: Is one solution here to hire more caseworkers and entrust more deci-sion-making to them? I was listening to an interview with you on NPR where you acknowledged that, hey, there's always been bias and problems with the work of caseworkers, but it does seem like at least they provide more flexibility and a human touch.

VE: I'm paraphrasing a good friend and political scientist, Joe Soss, who says that discretion is like energy: it's never created or destroyed, it's simply moved. One of the things we need to be keeping our eyes on is that these digital tools don't actually remove discretion, they just move it. In Indiana, they removed discretion from caseworkers and gave it to the private companies who built the system. It moved discretion from frontline casework, public workers, and gave it to IBM and ACS. And really we're tracking how these systems work within the context of the attack on public workers. Because often these sys-tems are rationalized as being aimed towards removing bias in the frontline of public service systems.

I'm also really really suspicious of this philosophy that says that human decision-making is somehow opaque and unknowable, and computerized deci-sion making is fair and open and transparent. Computerized decision-making is not as transparent as it seems–there're all sorts of hidden assumptions built into these systems. And on the other side, I actually believe that we can talk about human decision-making, we can address bias in these systems without removing discretion and deskilling frontline caseworkers. That can be part of people's professional development, for workers to talk through how they are coming to their decisions, and where they think their assumptions are coming from. I think frontline caseworkers have skills and knowledge that are really important to people getting access to the resources they need. I'm very suspicious of systems that are intended to remove discretion from the folks in the system who are the most working-class, the most female, and generally the most diverse of the whole public-services workforce.

JS: How do people organize against these systems, or at least try to critique them and secure their own rights?

VE: I think we are really legitimately trying to get our heads around that right now. I can give you an example from the book. In Indiana there was a successful mobilization of ordinary people against the system and that's why this ten-year contract got ended three years into the contract. That system did get rolled back. There was some really incredible on-the-ground organizing in Indiana that started with a process of town halls where people spoke openly about their experiences of trying to maintain their public services under this system. It really started with making sure that there was room for those voices of folks who were most impacted by these systems to share their experience. And that became a chain of town halls across the state, which launched a bunch of press tours where journalists talked to people who'd been impacted and created political pressure, which led to a successful bipartisan pushback against this system. So there's definitely models where folks have done this work in the past and have succeeded at it.

I would love to see more of this work where we're raising the voices of those folks who are most directly impacted by these systems. And to me, that would be folks on public assistance, unhoused people, folks interacting with the criminal-justice system, and folks who are interacting with the child welfare system. I think those are great places to start, because those are systems with low expectations that people's rights will be protected, so that's where some of the most abusive, invasive policing systems are found.

JS: Does it help at all, as you mention in the book, that these systems affect all of us in some sense? That seems like a place where we can make common cause, and while these systems affect the poor most acutely, it seems like we're all having data collected about ourselves and being subjected to systems of assessment and sorting.

VE: I start the book with a personal story about my partner being attacked and ending up needing some really extensive reconstructive surgery. We got through this incredibly difficult time together, but one of the most difficult moments for me was when our health insurance was suspended. I suspect that it was suspended because they were doing a fraud investigation of us and they suspended our benefits until they had completed that investigation. I'll never know, I'll never be able to confirm that, because the insurance company says there were just a couple of digits missing in the database.

The reason that I start with that personal story is to say that we don't all interact with these systems in the same way. My family wasn't often caught up in the Medicaid system and child protective services. There isn't that kind of digital scrutiny of my neighborhood on a day-to-day basis. We managed to prevail against this attempt to block us from health insurance, so we don't all experience these systems in the same way, but we do increasingly all come into contact with these systems. And part of my goal in writing the book is to help us see the way our experiences are mirrored in each other. I do think that's a place where we can start to build solidarity and alliance and power–recognizing that while we're not all the same, while we're not all *equally* impacted, we are *all* impacted.

And that's what happened in Indiana as well. One of the reasons why the organizing was so successful is that it started to affect folks' grandparents, who were receiving Medicaid in nursing homes, and that created a really broad and wide pushback against this system because it seemed to be affecting everyone. There are ways to see similarities in our experience, but, again, I think it's crucial to start with the people who face the short end of the stick, because they have the most information about the problems, and they're most invested in creating smart solutions. ●

A COMPENDIUM OF LAW ENFORCEMENT SURVEILLANCE TOOLS

By Edward F. Loomis

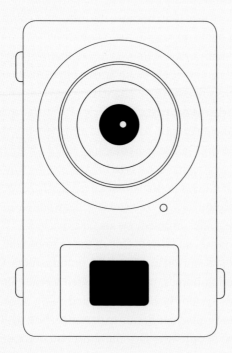

Body cameras provide a convenient means to capture and record both audio and video encounters with the public from an officer's first-person perspective. The most popular camera models are as light as 2.2 ounces and as small as 1.9" × 1.9" × 0.75", and deliver resolutions from 480 VGA to 1080 HD with viewing ranges up to 142 degrees. Some body cameras contain added features providing 32 GB of internal storage, GPS for geolocation tagging, and infrared LED lighting for night and low-light situations.

In September 2015, the U.S. Department of Justice launched a $23.2 million grant program to purchase body cameras ($19.3 million), training and technical support ($2 million), and studying the impact of their use in law enforcement jurisdictions within thirty-two states ($1.9 million). The presumed benefits of the program were that both the public and police might exercise better behavior with awareness that their encounters were being recorded. The cameras would serve as a reminder for officers present to adhere fully to departmental protocols, and the images and audio recordings would serve as evidentiary documentation in resolving complaints and court cases.

Body cameras surveil everything in the vicinity of the lens, capturing both audio and video—including innocent parties nearby who may be unaware they're being recorded. The captured video may include the date, time, and GPS coordinates of the recording. Some cameras deliver livestream communications to a remote server for live monitoring.

Body cameras have gained increased public acceptance since being deemed valuable in accountability disputes between cops and the public. However, due to fear of retaliation or fear of public exposure, there exists the concern that the use of body cameras may discourage bystanders from coming forward as credible witnesses to help assist with investigations.

The effectiveness of their use is somewhat mixed based on the results of two contrasting studies. The first yearlong study was conducted within the Rialto, Calif., Police Department from 2012 to February 2013. Use-of-force incidents occurred twice as frequently during patrol shifts where cameras were not worn as during those in which the cameras were used. For the Rialto PD, the use of body cameras reduced the use of force incidents for the year studied to 60 percent of what they had been in the previous twelve-month period. Also, there was an overall reduction in the number of complaints lodged against the police, dropping from twenty-eight filed during the year before the study to just three that year.

More recently, a study conducted by The Lab @ DC on the employment of body cameras by the Metropolitan Police Department of Washington, DC, during the eighteen-month period ending in December 2016 revealed no statistical difference caused by their use. The study found that the devices "have no effect on the measured behaviors, and the video footage they produce has no effect on judicial outcomes." The study posits that

▷ **A Compendium of Law Enforcement Surveillance Tools**

○ Facial Recognition Systems
○ GPS Trackers
○ License Plate Readers
○ Drones
● Body Cameras
○ Cell Tower Simulators
○ Parallel Construction

`MCS54` 0248

"perhaps neither the officer nor citizen involved in an interaction are actually aware of the camera, either due to attention being diverted elsewhere or desensitization over time to the presence of the cameras." The study concluded that it was unable to detect any statistical effect from the officers wearing body cameras. It further cautioned law enforcement agencies in environments similar to Washington, DC, that may be considering adopting body cameras not to "expect dramatic reductions in use of force or complaints, or other large-scale shifts in police behavior, solely from the deployment of this technology."

A November 2017 report produced jointly by Upturn, Inc. and the Leadership Conference on Civil and Human Rights examined police department policies on the use of body cameras. From the data collected in seventy-five major departments and jurisdictions across the country, the report concluded that police body camera policies vary widely.

The report examined the jurisdictions' use policies that were publicly available in order to determine whether or not they met the Conference's eight criteria of civil rights principles on body-worn cameras: policy availability, officer discretion, personal privacy, officer review, footage retention, footage misuse, footage access, and biometric use. Departments that met a simple majority of the criteria were those from Baltimore City, Md.; Cincinnati, Ohio; Montgomery County, Md.; and Parker, Colo., each having satisfied but five of the eight criteria.

The ratings were categorized into four classes: those fully satisfying the specific criteria, those partially satisfying the criteria, those in which the policy either did not address the issue or ran counter to the principles, and those where the department had not issued a publicly available policy on the issue. For the seventy-five departments and the eight issues examined, six hundred policy issue ratings were reported. 143 issues fully met the Conference's criteria, 126 issues partially satisfied the criteria, 303 failed to meet the criteria, and 28 ratings could not be determined because those departments' policies either were not made public for review or did not exist.

The 2017 study reported the following results for the eight criteria against the publicly available body camera policies of the seventy-five departments examined:

- Twenty-three departments did not post a policy on their website at all.

- Twenty-seven departments clearly described when officers must record, but did not require officers to provide concrete justifications for failing to record required events; six departments lacked any policy on when to record.

- Eight departments did not address personal privacy concerns of those being filmed.

- None of the seventy-five departments had a policy requiring officers to file an initial written report or statement before relevant footage was reviewed for all incidents; fifty-nine

departments had policies allowing or encouraging officers to view relevant footage before filing an initial written report or statement; and four departments had no specific policy regarding footage review by their officers.

- Sixty-one departments lacked any policies requiring the department to delete unflagged footage.

- Thirty-six departments did not expressly prohibit either footage tampering or unauthorized access.

- Sixty-four departments did not expressly allow individuals filing police misconduct complaints to view relevant footage.

- Sixty-eight departments imposed no

restrictions on use of biometric technologies to identify individuals in footage.

While policy improvements were noted by the 2017 study compared to results from the initial 2015 study, all police departments still have considerable room to improve protections of the First and Fourth Amendment rights of individuals whose images are captured. ●

The Postcards
We Send

Tips on Staying
Vigilant in the
Information Age

Soraya Okuda

Introduction

A heart dives into a Japanese character, a curly hiragana あ (*a*), and then into a cursive Arabic ي (*y*). Written sloping down at a diagonal, the pen pulls left to right, then reverses. Persian and English grammar, with made-up vocabulary inserted and retaining the original phonemes, compose the sentence.

Lomay. "Love you."

This was something I did as a kid. I loved making up phrases based on other languages and writing short, secret notes for myself. These were never notes of importance, but they were readable to me and only me, which made them special.

I imagined these notes were private: others didn't know how to unscramble the contents, and the rules that I based them on were disparate and obscure enough that an observer wouldn't want to put in the time to figure them out.

I knew other kids speaking Pig Latin, which was indiscernible to me until someone explained the *–ay* rules. Others, using cereal-box prizes and decoder rings, assigned letters to numbers, incrementing them slowly into scrambled language.

I wasn't special among kids. We all liked having something, written or spoken, that wasn't understandable to most people, something that required decryption and an understanding of our rules.

Sometimes we'd let other people in on them. For example, let's say I decided to share the key to my language with only one person: a good friend, moving across the country. The language was just for her and me.

When we saw each other for the last time before her big move, I'd share how to write the language. I'd hand her a key explaining what each character means, what the basic grammar is, and how to read something I sent. She'd hold tightly onto that paper legend, that key, before moving away. I'd also have the same key for myself, so that I could remember how to read her messages to me, and write something back, in our language. We'd keep our keys in a safe spot, tucked away in diaries protected by locks, hidden from our parents and siblings. Or maybe folded in lockets around our necks.

We'd send each other postcards in our secret language. The facts about the message—the type of paper and pens we used, the visible to and from, our locations, the little rips and tears from being handled along the way, the dates and carriers—would be known. But the content would remain gibberish to anyone else: encoded to both of us, readable by both of us, and visible but not understandable to others.

In many ways, our computers send information over the internet like postcards. Metadata—data about the message like our stated names, locations, services, and networks, the times sent and received and read and replied to, and perhaps even how it was handled along the way—is visible. By default, information sent online with plain text, as with a postcard, is also visible. However, we can protect the content of these messages, using encryption to render it "ciphertext" to others. When encrypted, the metadata is still visible, but the content is unreadable to others without the key. We can now rely on our computers' algorithms to use hard-to-crack encryption.

The key I shared with my friend for our secret language was symmetric: the same key used to scramble messages was used to unscramble them. However, what if my friend and I were hesitant about sharing the key in person?

What if my friend's sister snuck into her room, found the key as she was sleeping, copied it, and referred to it to read our messages? What if she forged my friend's handwriting, and pretended to be her in a letter? On the internet, this is not so far-fetched. We call this a "man-in-the-middle" attack. Someone intercepting a message can choose to look at the message, tamper (or not tamper) with it, and pass it along. The actual conversational participants may never know. Modern asymmetric encryption protects against this scenario.

Asymmetric encryption, or public key encryption, accommodates geographically distributed communications so we don't need to physically meet in person to share a single key for encrypting and decrypting information. It also provides a way to verify identities.

From HTTPS to VPNs, from cryptocurrencies to the Tor network, the principles of symmetric and asymmetric encryption are seen everywhere in modern encrypted communications.

Encryption has progressed significantly, but strong encryption isn't the default for some of our digital communications. For many people, the stakes are higher than a family member reading secret postcard messages to a friend.

That we send unencrypted "postcards" in many of our communications is, to put it lightly, a problem. The average person now needs to worry about credit card numbers and social security numbers being snatched by opportunistic thieves while on an unencrypted service. As we look for our friends through masses of chanting people at a public protest, our text messages and calls reveal a history of our contacts. Whistleblowers contacting journalists must worry about their employers watching their network communications, eager to punish any sign of disobedience. Journalists are concerned about businesses stalking them, organized crime threatening them with violence, their governments paying extra attention to their incoming and outgoing messages, hunting for sources. We cannot even trust many of our telecommunication networks, which not only profit from collecting data on our physical locations and purchasing habits but are themselves vulnerable to hacking and to government and law enforcement requests.

It is easy to feel a sense of nihilism about the state of security and privacy. Yet there is hope. Security is a process. Security is continually learning and improving. Security comes down to good hygiene.

In digital security workshops, we like to recall metaphors from public health and childhood education when discussing these topics. Remember when you were told to wash your hands with hot water and soap, remember when you were told to be wary of strangers, remember when you were taught to seek people with integrity and to avoid those who were dishonest. Eventually the following suggestions will come as second nature. Adopting even one or two of these suggested measures—like encryption in transit, using encryption at rest, and practicing basic security hygiene—will help your devices remain free from unauthorized access. Make yourself a nice cup of tea and take it one step at a time. You'll get there.

1.
Protecting Postcards
Using Encryption in Transit

Each unencrypted postcard carries metadata, such as your computer or phone's IP address and other unique identifiers. This postcard also carries the data, or content, of the communication. This postcard is being copied as a file at the hands of each computer (servers, network devices, and so on) and passed along until it reaches your friend.

And it's not just the computers passing along the message that can see it: this flow includes potentially thousands of computers on shared networks that pass along the message. For example, someone sharing your unsecured wi-fi connection on a plane or in a hotel or at a conference could intercept your activity, eager to see you checking your emails, sending pictures, or looking at a website. Imagine the fiber optic-cables undersea, on land, and in carrier facilities—the literal internet—that carry humankind's communications and store them, being intercepted by computers owned by the NSA, as Edward Snowden revealed.

These postcards, however, can be protected from prying eyes. There are a few types of encryption that do this, categorized under the broader umbrella of "encryption in transit"— encryption that protects your information while it's on its way from you to your recipient.

Encryption in transit should provide three main benefits. The content of the message is encrypted from prying eyes, providing privacy. Additionally, the integrity of the message is preserved, in that it's very difficult for someone to intercept, tamper with, and edit the message. Lastly, asymmetric encryption provides a means for identification.

Identification is one of the big themes in digital communications: how do we prove that we are who we say we are? For people who are targeted by man-in-the-middle attacks (such as activists, journalists, and human rights defenders), the ability to verify identities is an especially useful property of encryption in transit. If we use measures such as public key fingerprints, certificates, and digital signatures for verification, we can make it much harder for others to impersonate us, our friends, or a service.

Public Key ▼

-----BEGIN PUBLIC KEY-----

MIIBIjANBgkqhkiG9w0BAQEFAAOCAQ8AMIIBCgKCAQEA7/f9B2ijy0GdLDnEOZNlYE
rJWF7MXGg8ZM07MLj+/UdE/C8miNrk2PMMPezuznMWGp8fau175sdngwxX4+huk93O
X+6RWoRCy8uhlb0Q7TKrDbLkuXJzfPeziJVR3Q8s7HRNbMeY4qzAzJqkeoRfWLglq
lvDCH9NqcH5GusssaLo/HH2B1zLWvhulJI74UfawBnGMqguscJO3H0MFJSPtiIOy1
oh96jTfw12GmhQEt8zYh8enYeHL9hzQ1fsMX6Fg/4fU4Dd2hDO4VHG6m/dWpse999
jPi8dkfIRWX9a8P7EkXRbwAyHqZhqgAj1SSbfnpDFxNop/3pq5H9XIrl5xwIDAQAB

-----END PUBLIC KEY-----

Private Key ▼

-----BEGIN RSA PRIVATE KEY-----
Proc-Type: 4,ENCRYPTED
DEK-Info: DES-EDE3-CBC,5F599C6178FF36F2

xLffrsOaMQNVUuy+4CKEENCPxbdChBHAQeB4u0xEaO/AMsLogf5mCSQ05GuoZX+m+ft
OHPYfq8sWju9ikjmaHHe1s6BZWMRI2yh43qbYqTi3wu74zn05uNa+mcIiYip3xjB0ZZ
rynEm5+RC9coJzvPcc/deBAdU5iKteXXafhkmWvou4g9wM2BYglZSDdLyO1iG4G8LYZ
v45pp5U30JTHoW1NfewmzMXaaR5/KCoCgM5QV+eFaj6fRfD2UJoY3oY734zcmRa8jqx
+gu5g0T8En3o4KenvykOZBkG0JOxL3oPecG1W0YYA+b2j/HiR+Zs2/CEGCnE9pN3HRg
Ccd7HiXgknisxGDO5H0hyLjsMbVyAurqF0GXNOpLbTGJia91xBRbqeOJ00RCUSty+pf
KAVljyop/3ZyzMaGuZEXxkdiwPwrYQNnRe20YZUVOQVX12R/TZP0+6EPEJC53U7G6Py
35G8TWod8I3ZWGc2hmBd/SUmMRwrmyLCZLPbYxb3/6FKLLyhFuymddKcdoKhTsdwDw/
0rTrW9PValh3awTN3UAgY3ZACxqDd401LTvd3t50Gbki4U74HnqhyHXo6kLp3ycqaul
Vl1CReAlL6ZwzF1WnzBnJnN0oIGZ/61MfGZmSPPWTsIz/MDYXfLLbJAwLv1I7kJDR5
6kqoihnEzh7YMFRwM7KK65LHjn6wjcMRPX/SbT17nxXqXUpA7zfoEqUseZZS2VoK0l
15/WBQ4ZvQ6+Tiy+hW4dz8hGU5GsmF6QKEkJqhuyjzefyUnUaDKWSHDcGlQh3boOBY
A7ZSvMtw9s7F9Yx79omx/kyV5AalAVdFQDQTMoT8q59OHjNspSZfXPXito5XGum6xa
TseKNsQJYgM15w3yOfQE21Hq8XQUn7xTxmx77FCaZG39zixNeuWACOjuRhNn78UiOo
h1X6d9YQVtxh8NbhZhxq5QdUDpw8gFa+0iSyg/rt2Sww/+4a/vJdkbWXpb7iEwIc+b
Ra4tgumSsPnoOKJ6Y/ssYx2y3O1gTN7SC+u+FKUaOcK+1fzJ2TLSB4H4AanyVxKA7n
b+GU54Qx/znWw/fh1GOK4bJ1DU9SbGo2GHvVHDL8vrDimVy51hjEvtJ+hwgFvMF6T7
qjGfRWMTNFlwKtcqmTSWxGy8ergjvhN/NkP3e1cRcfVO2PeIbdITBqbtus7C4JNMiP0
MvVyA5e9QKgi/+6bDhL3ezkh8TyCVIKAFfwhA2b6UZuO9WeMd3iAAE+UAvU09hTuVd5
aqzQi712306jxTJoBeD3hhtgmm6b4F7w3EidA9F9LN4dsb2dFKrGS2eKjRuYLe/MoLh
6yIrBmbd//roJqFynrJ4HH4IBqmq3Kmh+sU0Oz00+Sarj7KRBAUhyoO1O2v2i4Er+8/
OVPyJtN72MFdB8JnsVoDQ8xVLKdihKZCB1C4bHDqj2NwLWMjKaVz35z5svcVydqWR6C
9hdAhutOE/fA/JpV7ixcUPG+L7mTJN7H78cIYRbTKZFmmJFKX8UKr+VFUynsVwr6wfG
tFwH0oK0tq6cED9g+qUY+Fuaun/+YYleBfIWcF9m47wLy7WHMm1wR29JD9XdzvoiTvK

-----END RSA PRIVATE KEY-----

△ Examples of public and private keys, created using the RSA algorithm.

Encryption in transit relies heavily on principles from asymmetric encryption. One well-known asymmetric encryption algorithm, RSA, relies on the principle that it is easy for computers to multiply two large prime numbers, but nearly impossible to untangle them from the resulting product. This algorithm has made it possible for us to exchange encrypted information with people in remote destinations without both having to know (and figure out a safe way to exchange) a key for decryption. Asymmetric encryption accomplishes this by having two keys instead of one: the public key, which someone uses to send you encrypted secret messages, and the private key, which you use to decrypt that message.

On the facing page is an example of a public key and private key pair, created using the RSA algorithm.

My friend would use my public key to send an encrypted message to just me. One's public key is shared widely on one's website, in a database, or in a message—metaphors fall short here, but it's kind of like an address to a locked mail box.

The private key is a file that holds the secret ingredients that created the public key. For RSA, these are the two large, randomly chosen prime numbers. The private key is what unlocks that which the public key locks, and must be protected and never shared. I likely wouldn't interact with my private key at all when decrypting my friend's message, as my computer is doing the work: decoding the secret language between her computer and mine. One's private key can be used to leave an unforgeable digital signature, like a personal seal, on the encrypted messages one sends, which a recipient can cross-check with the sender's public key to authenticate their identity.

If my friend and I were using encryption to write to each other, we would both have our own set of keys, and each other's public keys, saved to our computers.

If the private key is "lost" or shared, someone else can decrypt the messages that we went through great pains to keep private, and potentially impersonate us. For example, if the above public key and private key pair were my own, I would have already

▼ Tip No. 1
Use HTTPS whenever possible. Download HTTPS Everywhere to make your browsing more secure.

compromised my message security by sharing my private key beyond my own computer in an email to my editor, let alone publishing it. I'd need to generate a new keypair for secure communications, and notify my friends that my old keypair was compromised.

Deleting the private key (or losing access, such as being unable to transfer the private key file from a broken computer) can mean that we are no longer able to access information encrypted to that matched public key.

Two of the main methods of encryption in transit are transport-layer encryption and end-to-end encryption. Both methods use these key principles from asymmetric encryption, but with different results as to who can read these encrypted postcards.

TRANSPORT-LAYER ENCRYPTION

Let's say you are looking to join a free dating website from your phone or your computer. You type the website name into your browser and land on a page beginning with <http://>. Anyone along the path (someone eavesdropping on your wi-fi connection such as another patron at a coffee shop, or your internet service provider (ISP) or telecom, or someone on the website's network) can see these requests. By using simple software, they can see the password for the account you created, the email you registered it to, the people you matched with, your dating preferences, your gender, your sexual orientation, the messages between you and your matches, and so on.

HTTP (Hypertext Transfer Protocol) is used by your computer to exchange information with the servers hosting the website you are accessing; HTTPS is the secure version of that communication, often connoted by a little green lock next to the URL in your browser.

HTTPS provides transport-layer encryption, which encrypts information between the service (including the computers and administrators of that service) and you. Eavesdropping computers in the middle of your connection cannot see the pages you're looking at, the forms you fill out, the searches you make on the site, or the actions you complete. You are still, of course, putting your trust in the service itself, as it will see your information as entered.

Transport-layer encryption uses both asymmetric encryption and symmetric encryption: in other words, a whole lot of keys. When a website uses HTTPS, your browser and the website's server have a very fast set of interactions called "the handshake," in which they verify each other's identities using their public and private keys and begin an encrypted conversation.

The intermediary computers passing along these messages are still able to see metadata (that you are connecting to the web address) but nothing further.

For example, say I wanted to look for "surveillance self-defense encryption" using DuckDuckGo's search engine. DuckDuckGo is protected by HTTPS, so someone on my network could only see me access <https://duckduckgo.com/>, and not <https://duckduckgo.com/?q=surveillance+-self-defense+encryption&t=f-fab&ia=web>.

The good news is that HTTPS is becoming widespread and is easy to configure. Using it as much as possible helps you to be more secure in your habits. Some websites have an HTTPS version, but not as the default: if you'd like to use encryption on websites whenever it's available, you can download EFF's free privacy-protecting browser extension, HTTPS Everywhere. If you run your own website and would like to enable HTTPS to protect those visiting your website, you can get an HTTPS certificate for free from the certificate authority, Let's Encrypt.

Using HTTPS as much as possible is a great thing. However, just because a website uses HTTPS does not mean that the people who own that website or host additional code on the website are choosing to protect your data. It's important to note that, just because a service uses some encryption, it does not necessarily mean that the service values privacy for its users.

Below are a few examples to be aware of.

When you visit a webpage, parts of the page may come from computers

▼ **Tip No. 2**
Download Privacy Badger to prevent
third-party tracking across websites.

other than the original server of the website you asked to visit. When you are on a site, you'll likely see third-party ads, like and share buttons, comments sections, and other embedded images and code. In some instances, this third-party content can carry methods to uniquely identify your computer's browser and track your browsing habits, purchases, and interests as you move from HTTPS website to HTTPS website. To prevent this kind of third-party tracking, EFF has a free browser extension for the Chrome, Firefox, and Opera browsers called Privacy Badger.

Some web services might choose to collect and retain as much user information as possible. Let's say you are using an HTTPS-protected site for project management, chat services, social media, or email. Perhaps you've created a small, enclosed group within that service: it can be easy to feel that the communication is protected between you and the people in that group. The HTTPS service itself will have access to this transport-layer–encrypted set of communications, with a number of employees having access to these keys through corporate servers.

For those concerned about corporate surveillance, there is a risk that this HTTPS service may choose to parse and save the information for marketing or other purposes. And for those concerned about government surveillance, there's a risk of law enforcement requesting access to your user data. It's critical for companies to fight against these requests, and to tell users about government data requests, publish annual transparency reports, and require warrants for content. Unfortunately, users of these services are burdened with the effort of reviewing privacy policies and transparency reports to assess the protections a company might provide.

Transport-layer encryption, such as HTTPS, provides significant protection from opportunistic attacks. But what if you do not feel that you can trust even the service to protect your communications from law enforcement or from the corporation itself? What if you do not trust a company with even your metadata, or with the keys to decrypt your communications?

END-TO-END ENCRYPTION

Most web traffic is encrypted on its way to the server with HTTPS, but most email, chat, messaging, and streaming audio/video services are not encrypted end-to-end. This means that the company that provides you with a service—whether it's Google for Gmail, Microsoft for Skype, or Facebook for Messenger—is free to access your data as it flows through the service.

These troves of data are visible to a malicious employee with access controls, such as a disgruntled systems administrator, and can be handed over when requested by law enforcement and governments.

We now have services that offer "end-to-end encryption" to solve that problem.

End-to-end encryption is a type of encryption in transit that encrypts the communication to be accessible by *just* your device and your friend's device. All the services passing along your message, including the app or company providing the end-to-end encrypted tool itself, cannot decrypt your message. Though, the intermediary service still sees all the metadata, such as the fact that you are using end-to-end encryption and whom you contact when. End-to-end encryption is designed so that your decryption keys are yours alone, and your friend's decryption keys are theirs alone. The onus falls on you and your friend to protect your thread of messages—for that reason, using end-to-end encryption will make you more aware of using public and private keys.

Though it's perhaps not the most practical use case, end-to-end encrypted email is one of the more illustrative examples of how asymmetric encryption concepts are implemented. Let's say my friend and I use an end-to-end encrypted email software from the early '90s called PGP. Both my friend and I have downloaded PGP using software like Mailvelope, which can be used on top of our regular email—say, Gmail. Email in Gmail is encrypted in transit using HTTPS and STARTTLS, protecting it from eavesdroppers. However, let's say that my friend and I don't feel comfortable with Google having access to

▼ **Tip No. 3**
Message with end-to-end
encrypted tools like Signal
to safeguard information
between you and your friends.

our email contents. I'd write my secret message in the encryption software inside of my email to her, then encrypt it by addressing it to her public key. Once encrypted, the body of the email appears as garbled nonsense.

I'd add a boring subject line—the only thing not encrypted—address the email to her Gmail address, and send it off. When she received it, it would similarly appear as nonsense until she used her PGP software (which had her private key saved in it) to decrypt the message: "I miss you, friend."

End-to-end encryption is now possible for messaging, video calls, phone calls, and sending files. People who deal with particularly sensitive data, such as financial documents, medical information, client data, and photos or videos that others may want, now have easy-to-use options for protecting the contents of their messages as they send them.

Signal, a smartphone app by Open Whisper Systems, is an example of a free, open-source tool that uses end-to-end encryption by default. It can be used over wi-fi and internationally. And unlike end-to-end

encrypted email, where you have the single private key that grants access to decrypting your whole email inbox, a cool feature of Signal is that each message is encrypted separately using a new symmetric key, which gives your conversations a property called "forward secrecy." This means that if someone gets the key to decrypt one of your messages, they won't be able to decrypt the rest of them. More and more software developers are adopting the underlying encryption protocols and features from Signal; see for instance WhatsApp's switch to end-to-end encryption in 2016.

In transport-layer encryption, your browser and the website's server automatically check each other's public key information to verify that they are who they say they are and to decrease the likelihood of a man-in-the-middle attack. With end-to-end encryption, this identification process is not automatic: the responsibility is up to you and your friend to verify each other's fingerprints.

For those who are particularly concerned about preserving the privacy

▼ Tip No. 4
Check the public key fingerprint of your correspondent to ensure they are who they say they are.

▼ Tip No. 5
Regularly back up your data—and check that your computer is backing up encrypted chats in ciphertext, not plaintext.

of their messages—like journalists and sources, or service providers and clients—checking public key fingerprints is an especially important feature. Checking public key fingerprints means reading each number and letter from a string and confirming that it's the same as what's displayed on your friend's device.

Some of these services can be connected across devices, for example on your phone and your desktop computer. Keeping end-to-end encrypted chats on more than one device may seem like an appealing option. However, as the private key for end-to-end encryption is a sensitive file, storing your private key on more than one device can be a significant security decision. A caution: if you're not mindful of settings for backups, some services may back up end-to-end encrypted chats as plaintext, unencrypted, on corporate servers somewhere such as to Apple iCloud.

Using encryption in transit, like end-to-end encryption, protects the message as it travels. But what about when it reaches a device and is stored?

2.
Digital Homes

As we use our devices, they begin to construct another version of our homes. Our searches, reflecting our concerns and general interests, are the books filling our shelves. Our files, reflecting our work and our financial information, are our journals. Our photos and messages are our love notes to family and friends, reflecting our connections to other humans, the things we hold dear, the way we relate to each other, our insecurities.

Just as our homes can be robbed, our devices can have valuable information lost or snatched away. However, there are many things we can do to protect this information, including but not limited to encrypting the data at rest.

PROTECTIONS FOR WARDING OFF MALICIOUS SPIRITS

When a device is no longer in your control—a computer robbed from an office, or a phone that downloaded malware—it can still be protected.

If you value your information, and if restoring it is a priority, regularly back up your data. In case your device fails

▼ **Tip No. 6**
Regularly update your software to patch security holes.

▼ **Tip No. 7**
Don't plug your phone into free ports without a "USB condom" to avoid incurring malware.

and must be replaced or repaired, you can have a copy to return to. As mentioned, be mindful of backup settings (in particular, make sure your encrypted messages aren't backed up in plaintext to a remote server like iCloud).

Another important measure for protecting your device is regularly updating your software, including mobile apps. Software can be thought of as infrastructure, constantly built upon. Sometimes holes are found in the operating system, like the foundation of your device, that compromise its stability. These holes are like broken windows on your device, known by a network of thieves looking out for people just like you. Imagine someone looking through your computer's webcam or listening to your phone's microphone without your knowledge. We can often prevent this kind of breach by installing updates promptly and regularly and avoiding operating systems and software that are no longer receiving updates. While it feels like an inconvenience to run the update, it protects the overall safety of your system.

Sometimes, we accidentally welcome these vulnerabilities in. Malware

is malicious software, often developed with the intent of extracting sensitive data from your machine and sending it to someone else's computer.

Across cultures, we have myths of gilded, beautiful boxes that aren't meant to be opened. Whether it is Pandora's box or the punishing trick that offers a box of presumed treasures in Japanese folklore like *Shitakiri Suzume*, curiosity overcomes us, we open these boxes, and demons stream out. Malware can be spread in many forms, often punishing our curiosity or our eagerness for gifts. For instance, be suspicious of plugging your phone into free electrical charging ports in airplane seats, on college campuses, and on city streets. Unless your device plug has a "USB Condom" on (a USB cable or adapter that limits access to data pins when your phone is connected, keeping only the charging port open), connecting to one of these charging stations has the risk of putting malware on your device. Likewise, be wary of free CDs, free USB drives, and other people asking to plug in to your device. There are stories of hackers and spies leaving USB sticks in

▼ **Tip No. 8**
Watch out for phishing messages sent to your computer or phone. Be mindful of what links and files you click on, as well as what files you download.

▼ **Tip No. 9**
If you receive a suspicious-feeling message from someone you know, double-check in person or by phone that they really sent it.

public places with the hope of targeted individuals taking one and unsuspectingly plugging it into their computers.

Extraction of valuable data can also be done though social engineering. Just as a thief can dress up as a repairman and (if unquestioned) enter your apartment building, others use tricks of impersonation to gain access and trust, as with phishing—a type of scam where a user is tricked into revealing personal or confidential information. One common type of phishing is sending tailored emails, texts, or social-media messages which look to be genuine. The phisher promises to help us if we click a link or go to a legitimate-looking webpage (often off by a single character or an obscured link), or download a file (for example, something posing as a .pdf, .doc, or, more suspiciously, .jar). Phishing also can take the form of a site offering software downloads or apps.

We click. Then malware is surreptitiously downloaded onto our machine. At best, it can just be adware—intrusive software designed to sell something—slowing down our machines significantly. At worst, it is able to undermine the measures we took to encrypt our communications, such as by recording what we type, what we see on our devices, our files, and what our camera sees. A particularly troublesome class of malware is ransomware, where a malicious actor can gain access to our machine, take control of it, encrypt all the information on it, and demand money for the decryption key. In 2017, ransomware known as WannaCry attacked more than two hundred thousand computers around the world. It was particularly damaging to the National Health Service in England, obstructing care for patients and forcing hospital staff to address a different kind of sickness: a computer virus.

It can be challenging to detect malware on our devices, but good anti-virus software can do some of the work of checking files with known malware, and operating system–provided tools like activity monitors or task monitors can check whether our computers are using energy on abnormal tasks. Unfortunately, the best way to address malware is to not get it in the first place: regularly updating software and being careful to avoid phishing attempts. If our devices do get malware, the solution often requires wiping the device and restoring it to a backup version. When hit with particularly persistent malware, we may need to get a new device completely.

Or perhaps, as is common, scammers socially engineer you through artisanal, tailored phishing attempts known as spear-phishing. Even if they don't have the intent to install malware on your device, a similarly bad outcome may be their gaining access to a single account for a website. By misplacing our trust in an impersonated service, we grant the attacker access to our account that holds the information they desire.

John Podesta, the chairman of Hillary Clinton's 2016 presidential campaign, clicked on a phishing link claiming to be a genuine Gmail site, enabling access for attackers to see the Democratic National Committee's campaign emails. It can happen to anyone and is one of the cheapest and most popular ways to compromise a device or account.

Humans verify other humans by the cadences of our voices, visual markers, the ways we walk, the words we choose. Humans verify computers and servers by checking carefully for spelling or by double-checking that a connection is encrypted correctly to the computer we think we are connecting to. Unfortunately, we're susceptible to ignoring what otherwise might raise our suspicions: we are willing to overlook intruders dressed as repairmen entering our building, and websites that have the HTTPS green lock and seem professionally produced.

Some ways to mitigate phishing are to verify with your friend in another channel (like in person or over a phone call) that they really sent you that suspicious-feeling email before you open it. Be aware that malicious actors purchase domains that look like genuine domains but whose URLs are one character off. Be scrupulous and type in links yourself, or even better have the genuine site bookmarked rather than click within an email or message. If the email contains an attachment that you weren't expecting, use a cloud-based in-browser viewer like Google Drive or Dropbox's preview functions. Better to open the file on someone else's computer (Google's or Dropbox's or something called a

▼ Tip No. 10
Set up notifications for logins to
your accounts to keep an eye out for
stolen credentials.

▼ Tip No. 11
Use full-disk encryption on your
devices and backups.

virtual machine) than to risk your own.
You can also monitor for strange activity and logins from your accounts by
signing up for account notifications.

And so, unfortunately, we must be
vigilant to keep our devices clean and
protect them from malware, to back
them up, and to update them regularly.

The last major way to protect the
data on our devices is to use full-disk
encryption (a type of encryption at
rest), scrambling it into unreadable
gibberish to everyone but ourselves.
Full-disk encryption generally does
not protect your device against malware. However, it does protect your
device's data from physical access.
Imagine a thief finding your pow-
ered-off computer, turning it on, and
being frustrated that they can't access
the information inside at all. Many
computers support device encryption,
such as FileVault for Apple computers
and BitLocker for Windows devices.
It is wise to set aside time for your
device to fully encrypt these masses
of files, and to back up your data
before enabling full-disk encryption.
(You can encrypt these backups, too.)

Many modern smartphones enable
full-disk encryption by default as
long as there is a password. However, having a password and being
encrypted by default are distinct
processes, and just because you
have a password does not mean
your device is encrypted. Be sure to
check whether full-disk encryption is
something that needs to be manually
turned on for your device, in addition
to setting a password.

WHICH BRINGS US TO PASSWORDS

Use any of these?

```
12345678, Password, Monkey,
Letmein, Dragon, Iloveyou
```

Computers verify other computers'
identities by matching public key files to
public key fingerprints, digital certificates, and digital signatures. Computers
play even stranger identity games with
humans, constantly asking us to prove
that we are who we say we are. Are we
proven by our biology? Are we what we
know? Can we prove who we are by a
sacred object that we carry with us?

Entire systems are built on these questions of authentication.

As our world becomes more and more like science fiction, we have systems that can be unlocked by our very bodies. We prove who we are to our phone by scanning our thumbprint or our face. These methods have their own shortcomings. For example, using a thumbprint to break into a phone is easier than we might imagine: thumbprints can be forged with printers and law enforcement officers visit morgues to unlock devices. Now we're forced to ask questions like how biometric information is protected by law, especially when used to unlock devices. Many law enforcement officers treat devices protected by biometrics as though they're not protected at all.

Then there's the method of verifying who we are with knowledge that must come up from our minds. In *One Thousand and One Nights*, entry to a cave requires someone to give a passphrase (a phrase of two or more words). The entrance opens only to someone who knows the passphrase: "open sesame."

Our phones, devices, and online services demand PINs and passwords. They ask us to answer security questions, which are often tied to publicly searchable information like our mother's maiden name, the schools we attended, and our past addresses. We construct passwords based on our own interests, our favorite songs, short words like *monkey* and maybe a *!* and a *1* thrown in, things about our loved ones. We're likely to use the same password across services because it's hard to remember them all. We're likely to use birthdates, we're likely to use our home address, we're likely to use things that we've already memorized.

In other words, the way we choose our passwords is predictable. And therefore exploitable.

Every time there's a breach where "millions of emails and passwords have been stolen," you can expect that malicious actors are trying to see how many people are using the same password across different services. Because many people reuse passwords, hackers will gain access to critical services with financial or other valuable information inside.

There is also the threat of those who know us well: if someone actually

▼ **Tip No. 12**
Use unique, long, random
passwords across services.

knows enough about you or is able to recall things you've written or said in the past, they may be able to just guess what kind of password you'd make. Even my made-up secret language would be easy to crack for someone who knew the languages I did, or who knew me well.

There are entire lists derived from dictionaries, song lyrics, poems, and pop culture references, which are then used by computers that can, within fractions of a second, try each password to log into a service. The shorter the password, and simpler the combination, the more likely it is that a computer will guess it correctly or that someone malicious will gain access. In one 2011 study of six million passwords by Mark Burnett, 91 percent of passwords were from the thousand most common passwords.

The interplay of these problems at scale necessitates our making passwords and security questions that are hard for humans to guess, as well as hard for computers to guess.

A hard-to-crack password is long, unique, and random. By long, I mean the length of at least a few words. By unique, I mean using completely

novel passwords for each service (and not merely a variation of each). By random, I mean there's no connection of each of these words, numbers, or symbols to each other, and that they're ideally randomly generated.

It's up to you how you retain a list of long, unique, and random passwords. Some people like to do this in a notebook that they zealously guard. For the rest of us, there are password managers, like LastPass or 1Password. A password manager is software that encrypts a database of your passwords, security questions, and other sensitive information, and is protected by a master password.

You might be wondering what kind of password protects a database protecting many, many other bits of sensitive information. You might want to use a passphrase, which can be three to eight random words, to protect this database. A random passphrase has the benefit of being hard for computers to guess (it has high entropy, and therefore too many possible combinations to crack) and hard for humans to guess. There's also the benefit of memory: it is easier for humans to

▼ **Tip No. 13**
Create a strong passphrase for your password manager using Diceware.

▼ **Tip No. 14**
Use two-factor authentication to add additional protection from phishing to your services.

construct a story around these random words and remember passphrases. For example, you can come up with a story for the five-word passphrase "wooded slander screen champagne serpent" as a mnemonic device: "In the wooded area, someone speaks slander on the phone. They turn on their television screen, pouring themselves champagne as they watch a serpent."

For those who are really into entropy and interested in generating strong passphrases, one of the recommended methods is diceware: a technique of rolling five dice to generate a five-digit number, then finding the corresponding word match from a sufficiently long list of unrelated words, until a passphrase of appropriate length is created. For example, using five dice, I roll 4-1-5-2-3. I look up 41523 on EFF's long wordlist and find its corresponding match: *mummify*. I roll four more times: 3-1-5-6-2, 6-2-5-4-1, 2-3-1-2-3, 3-3-3-3-3. My diceware passphrase is "mummify fructose tribunal despise handgrip."

Assuming you've done all the above, you're in good shape: you've managed to avoid suspicious USBs and devices,

you don't let anyone plug anything into your computer, you've updated your software, you've used random passwords across your accounts, you've been careful. But let's say someone eavesdrops on you sharing a password. Let's say you are spear-phished.

There is one major thing you can do, which at least protects your individual accounts in the unfortunate event that someone obtains your password. Like a talisman or sacred object that mythological characters carry with them so that others know they are who they say they are, computers can rely on another form of authentication (in addition to passwords) before granting access to your treasured account. The last measure of authentication is something you have, an object you carry with you.

Two-factor authentication relies on the principle of "something you know, something you have." The something you know is your password, whereas the something you have is a physical device or object of some sort that can be used for identity verification. Many social-media and banking services offer two-factor authentication, and it is one of the most effective things you can do to protect your accounts. Why? Because it's hard to steal a password remotely *and* steal a physical object from someone. This requires the user of two-factor authentication to be mindful to not lose their object.

Two-factor authentication is often in the form of a security token that fits into a USB port, like a Yubikey, which fits on a keyring. After entering your password, you are expected to use your second factor: tap the little flat button on the token, which then grants access to your account.

Or the sacred object can be a phone or tablet. You can download an authentication app, like Authenticator, which has time-based access codes that cycle every thirty seconds. (There's also the option of receiving an SMS, but as mentioned in the post-card section SMS is not encrypted and can be accessed by eavesdroppers.)

A good protective measure, if you are someone who is likely to lose their physical object or have it stolen, is writing down backup codes—one-time use codes that you can use if your primary two-factor authentication fails, or if you've lost your phone or token.

Developing good security habits takes time, but data hygiene for individuals and communities starts with these tips.

Things change, things fall apart. Advice can quickly become dated by new considerations.

To stay updated on advice, you can follow a blog called Deeplinks, which features analysis from EFF as the security and privacy landscape changes. If you're interested in more digital security advice—such as learning about threat modeling, pseudonymity, Tor, VPNs, and more advanced scenarios—check out EFF's Surveillance Self-Defense project. If you're interested in teaching your community about digital security, you can use EFF's Security Education Companion resource.

WHOM CAN WE TRUST TO PROTECT OUR DATA?

Our devices can be made more trustworthy through things like full-disk encryption and password protection.

Our accounts can be configured to require more verification, such as through two-factor authentication. We can communicate securely with services through transport-layer encryption like HTTPS. We can communicate with our friends in a more trustworthy way though end-to-end encryption tools like Signal. We can take meaningful steps to protect our data and to protect the data of our friends.

Trust is hard-earned.

The others that keep our data, like corporations and governments, must also protect what they have of us. They, too, must promise to encrypt our data at rest, keeping our information safe on their computers when it is collected. They, too, must use strong passwords and treat these protections seriously. They need to provide the most basic protections, such as transport-layer security and a mandate to not collect more information than we willingly give them. We should encourage and applaud services that provide end-to-end encryption.

For those of us who are privileged enough to do so, there's also an ethical responsibility to evangelize and liberally use encryption, as it provides greater protection for those who really need it.

The protections are twofold. There's a ripple effect of data mining: corporations and governments may look for your conversations in order to find the conversations of your friend. By using encryption, you help protect the information of those you communicate with. Additionally, you're helping to normalize the practice of using encryption in the first place: the more people use encryption, the more normalized the metadata of using encryption becomes. By increasing our own usage of encryption, we are able to help provide blanket protections for people who are targeted, like whistleblowers and journalists, who may otherwise be singled out or thought suspicious for their use of it.

A child might be asked, "Why does your diary have a lock? What are you trying to hide?" The principle is not that there is something to hide, but that, with the lock, we can be freer in how we feel and how we interact. That it feels good to have something private. That it feels good to write in a language of your own. ●

▽ ESSAY   A digital-rights activist calls for us to stop reacting and start acting.

0275

A More Visionary Movement

Thenmozhi ⊙ Soundararajan

S everal months ago I had a heartbreaking conversation with a friend who was working on the issues of data collection and the Dreamers. She was concerned that data volunteered by Dreamers who had registered with the United States government in order to get temporary protected legal status could now be weaponized by a Trump administration determined to remove all immigrants. The data she was concerned about included information about not only the Dreamers but all the vulnerable undocumented people in their families as well. This data, if weaponized by government agencies like ICE, could have traumatic impacts on the community when used for raids, deportation hearings, and worse. In the wake of this uncertainty she was searching for answers and had reached out. She asked me with gravity, "Is there any way that we can demand that the government destroy our data?"

And, of course, I didn't know how to respond.

How do you explain that, beyond the problem of the government having an immigrant's A-file, or Alien File, containing all of an immigrant's records, the United States government collects intimate details of our lives through its half-dozen surveillance agencies and often stores them in databases accessible to both government officials and government contractors? How do you explain that a request to delete something doesn't always mean we can guarantee a deletion? For, once data is collected, it can most certainly travel.

And ultimately, while I did eventually explain all this to my friend, what I realized in the poignancy of her question is that most Americans are like her. In my work at Equality Labs I have the privilege of working on issues of urgent movement security. In this role I coordinated a massive, national rapid-response effort after the election to support hundreds of grassroots community-of-color organizations that were grappling with very real concerns about increased surveillance and the rise in digital attacks by both the state and crowdsourced white-supremacist vigilantes.

As we worked with people in their most vulnerable hour, I saw many activists blame themselves for not knowing how to use their devices better, as opposed to blaming the larger failure of the government to protect their rights.

Even fewer viewed this specific surveillance within a larger understanding of the history of surveillance in our country.

It was then that I realized our people simply do not have the literacy they need to understand the violent systems of surveillance and their longer-term implications. They're instead caught in the real-world consequences of these systems wreaking havoc on their lives right now. From infants who are tagged as gang members in Los Angeles police databases to Muslims who are being targeted by Countering Violent Extremism programs to black activists who are being labeled Black Identity Extremists by the FBI for their work in Black Lives Matter, all of our communities are under attack.

These systems of mass surveillance were supposedly created in the name of making Americans safe from the bogeyman of the moment. Be they terrorists, immigrants, or gang members, these categories are built from implied racist stereotypes, used to frighten Americans into accepting these systems. This acceptance protects the government from demands for transparency and information about the longer-term impacts of invasive surveillance operations on the fabric of our society. It was never about informed consent. It was instead coercion built on a vicious and ahistorical narrative that presumes surveillance is only a technological problem.

Without a holistic understanding of these issues, our society processes the experience of mass surveillance piecemeal. In turn our movements create campaigns that are reactive and limited in scope because we focus on the symptoms, not the root causes, of mass surveillance.

The heart of our challenge as a movement fighting for internet freedom and privacy is this: we need to stop framing the problem of surveillance as simply a technological problem—it is instead a problem of state violence. We have to reframe the question of surveillance and privacy rights as one rooted in our larger battles against structural racism. Surveillance as a tactic of state violence has always been used against vulnerable communities in the United States. We have to use both history and the centered experiences of communities who are on the front lines of this violence as keys to understanding how we fight this dangerous threat to our democracy. To do this we need to create communities that are aware of the ways the United States has always surveilled black, brown, and indigenous communities. We can use this history to inform our communities of their historical resilience—a powerful foundation on which to gather strength, empower, and protect ourselves in the fight for our rights.

The time has come to reframe the problem of surveillance by better understanding its origin story in American history. We cannot begin with the presumption that there was ever such a thing as a golden age of privacy, for the Bill of Rights did not ensure privacy for all peoples in the United States but rather only for white, landed, cis men. Black, brown, indigenous, femme, and queer bodies have always been fair game for invasions of privacy by the state and private business.

The history of the United States is the history of the control of black and brown bodies. From the lantern laws of the 1700s to the Chinese Exclusion Act, which required the registration of immigrants for the first time; to COINTELPRO, which ruthlessly surveilled, harassed, and even murdered domestic organizers from the Black Panthers, American Indian Movement, Young Lords, and Brown Berets; to the NSEERS Muslim registration database, this is a violent history of technological innovation being used to bolster the bureaucracy of a white-supremacist American state. For the growth of our nation required the growth of its infrastructure to control, track, and surveil black, brown, and indigenous bodies.

In addition to naming this painful legacy, we must also move our understanding of surveillance beyond a purely libertarian point of view and examine this problem in the context of power.

Mass surveillance is an equity issue and it cuts across the landscapes of race, class, and gender. To address this we must use an intersectional lens in how we help our networks practice digital security and in how we create visionary policies of resistance.

Empowering communities of color when it comes to the issues of mass surveillance requires changing our entire approach. We must move from reactive, rapid-response models of managing surveillance to visionary models of

Spike in Hate Crimes
Around the 2016 Election

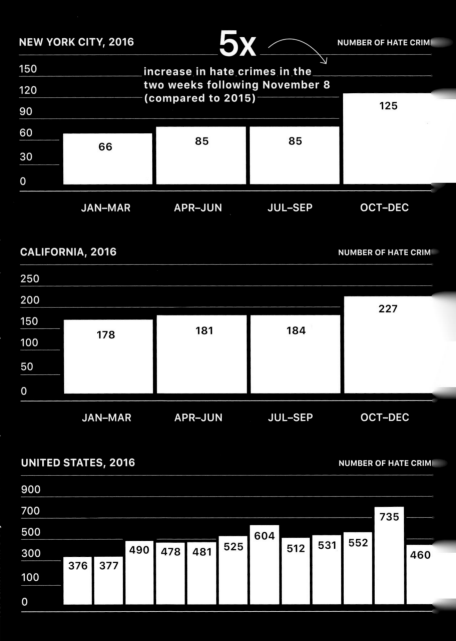

NEW YORK CITY, 2016   **5x**   NUMBER OF HATE CRIM

increase in hate crimes in the
two weeks following November 8
(compared to 2015)

	JAN–MAR	APR–JUN	JUL–SEP	OCT–DEC
	66	85	85	125

CALIFORNIA, 2016   NUMBER OF HATE CRIM

	JAN–MAR	APR–JUN	JUL–SEP	OCT–DEC
	178	181	184	227

UNITED STATES, 2016   NUMBER OF HATE CRIM

376, 377, 490, 478, 481, 525, 604, 512, 531, 552, 735, 460

data stewardship—models that are part of larger movements envisioning the possibilities of community-driven tech that centers cooperation, not capital.

This means sitting down with folks beyond digital security workshops and asking big-picture questions: What is our data? What does it represent about us? And, if we lived in a just world, what role would we want trust, sovereignty, and collaborative economics to play in our relationship with that data?

Conversations like this might feel existential in a time of urgency. But these are the visioning questions that help us think outside the frame of oppression and open us to newer game-changing possibilities. These are also the questions that help to create a culture of affirmative consent around data-driven economies, motivated not by capital but by community stewardship. These are not pipe dreams; they are the platforms for visionary resistance. For it is much easier to get people to fight for something than to get them to fight against something.

Visionary responses also help our folks overcome the inevitability discourse that accompanies much of the dialogue around surveillance and instead fight back strategically. This discourse presumes that surveillance is inevitable and that there is nothing we can do to fight it. And many organizers struggle against the idea of giving up when it comes to mass surveillance. Some internalize this attitude so much that even adopting simple prevention measures feels pointless. The inevitability discourse lends many digital security trainings a tone of pessimism and doom. That is primarily because our frame of reference is anemic. Rapid-response methods of fighting surveillance require us to always be in crisis. They presume we can't fundamentally control the ever-shifting impunity of tech companies and the state that's complicit with them.

But I believe we can and must do more. Moving to a visionary strategy around surveillance activism is more crucial now than ever. We are in a historic moment. For beyond the crisis of the Trump administration and its attacks on all of our privacy rights, we are also on the precipice of a new era of economics brought on by the Fourth Industrial Revolution. This technical revolution runs on data and will cement surveillance as the backbone of our global economies. The fact that such a significant shift is happening at a time when there is so little state protection for our privacy rights is significant. It means that the current level of surveillance enacted upon everyday citizens will scale in a way we have never seen before. Unless we engage now.

Few people in digital security or racial justice spaces talk about the impact of the Fourth Industrial Revolution on historically marginalized communities. But the convergence of artificial intelligence, robotics, the Internet of Things, autonomous vehicles, 3D printing, quantum computing, and nanotechnology is fusing the physical, digital, and biological worlds, and impacting all disciplines, economies, and industries. Like previous industrial revolutions, it redefines resources, supply chains, workers, and thus our communities—whether we like it or not.

In this new industrial age, data is the new oil. This analogy has all of the terrible but appropriate associations of extraction politics where, instead of destroying the earth, it is into our very existential selves that we are drilling for capital.

AI-related algorithms are strengthened and further built upon by the enormous computing power and loads of data accessible today. This makes surveillance the very foundation of the new information economy. Given this critical juncture in history, we need to have a strong movement built around an analysis of racial equity. Communities of color stand to lose more than just their privacy rights; they stand to lose their very stake in the future.

This is why a visionary approach to surveillance management must be rooted in history while also radically reimagining what is possible for the future. A killer app like Signal is not going to solve a broken technological ecosystem and an even worse policy framework. Only innovation in the ways we think about both the tech and the movement itself can solve it.

If you don't believe me, look at the ambition of the white-supremacy movement. When kicked off of major tech platforms for hate speech, these technologists simply created their own ecosystem with alternatives to Facebook (wrongthink) and Twitter (gab.ai) and even a funding platform to replace Patreon (hatreon).

The white supremacists of today do not limit themselves to reactive strategies. They instead innovate to develop power and work toward strategies that allow them to build the world they want—one in their own image. They are not fighting back; they are world-building.

If we don't want to live in their world then we, too, must scale our ambitions beyond reacting to our opponents, and instead build technology that reflects our values and—most critically—the world we want.

We need a new generation of organizer-technologists to not only help hold the line around crisis but also spearhead technology and policy development for new models of data management, community sovereignty, education, and regulation. The heart of so many of our problems is a flawed economic system based on the pursuit of capital above all else. We must let visionary conversations about the society we want, rather than a drive for capital, shape our tech.

There is so much more to say about this. But it is my hope that this can be a call to action to not give in to the dread. We have so much creativity and dreaming potential in our communities and, now more than ever, our radical imaginations are our greatest weapons.

Crisis can be a doorway to many opportunities. It is my hope that we can move beyond simply responding to the crisis and dream bigger. We are the future and the ones we have been waiting for. As the anxiety of mass surveillance and this presidency attempts to shut down our sense of possibility, one of the most radical acts we can take is to hope boldly. ●

A COMPENDIUM OF LAW ENFORCEMENT SURVEILLANCE TOOLS

By Edward F. Loomis

○ Facial Recognition Systems

○ GPS Trackers

○ License Plate Readers

○ Drones

○ Body Cameras

● **Cell Tower Simulators**

○ Parallel Construction

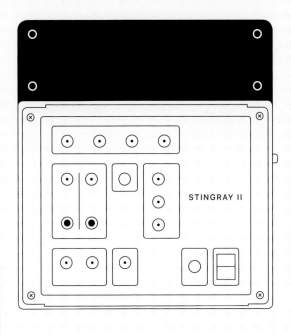

STINGRAY II

Cell tower simulator technology—commonly referred to by their Harris Corporation product names, Stingray and Hailstorm, or their Digital Receiver Technology product name, Dirtbox—was originally developed for the U.S. intelligence community and military. The technology was exhibited to the law enforcement community at the 1991 National Technical Investigators' Association annual conference, and law enforcement use first appeared in case law in 1995.

Since Department of Homeland Security funding became available to state and local police jurisdictions for supporting the counterterrorism mission, many departments across the country have acquired cell tower simulators. Use of the devices has since expanded beyond counterterrorism to service virtually all forms of criminal investigations.

Cell tower simulators mimic the behavior of standard cell towers by

▷ **A Compendium of Law Enforcement Surveillance Tools**

○ Facial Recognition Systems
○ GPS Trackers
○ License Plate Readers
○ Drones
○ Body Cameras
● Cell Tower Simulators
○ Parallel Construction

MCS54 0284

capturing the location and unique International Mobile Subscriber Identity (IMSI) of all cell phones within the vicinity of the equipment. An IMSI is a unique identification number stored in a cell phone that is used to identify the subscriber for network access privileges. The number has three components—a mobile country code, a mobile network code, and a mobile subscriber identification number.

Cell tower simulators work by broadcasting a stronger pilot signal than those of nearby operational service provider cell towers, thus activating connections from unsuspecting users' mobile phones. Some simulators capture additional information from the forced connections, such as metadata—including the phone numbers, time, and duration of the calls—and the content of text messages and websites visited. The equipment may be mounted on an airplane, helicopter, squad car, or drone for maximum mobility and stealth.

A November 25, 2017, Associated Press article revealed that "at least 72 state and local law enforcement departments in 24 states plus 13 federal agencies use the devices, but further details are hard to come by because the departments that use them must take the unusual step of signing nondisclosure agreements overseen by the FBI." The nondisclosure agreements are required in order to shield the devices' use and capabilities from the public and preclude criminals from developing countermeasures.

To illustrate the extremes to which the nondisclosure applies, consider the case of Tadrae McKenzie, who in 2012 robbed a marijuana dealer of $130 worth of pot at a Taco Bell in Tallahassee, Fla., using a BB gun. Florida law treats such crimes as felony robbery with a deadly weapon, punishable by a minimum sentence of four years in prison. Surprisingly, though, McKenzie was given a sentence of six months probation after pleading guilty to a second-degree misdemeanor. The lenient deal was prompted by evidence McKenzie's defense team uncovered before the trial that law enforcement had used a secret Stingray surveillance tool to investigate the crime. The judge ordered police to exhibit the Stingray and its data to McKenzie's attorneys, but they refused due to the nondisclosure agreement with the FBI.

When interviewed by *Popular Science* in 2014, ACLU attorney Nathan Wessler said, "You can imagine quite sensitive information that the location of someone's phone can reveal. You can tell it was my phone that was at the casino until 2 a.m., drove out to the brothel at 4 a.m., and then back to the casino at 6 a.m. Or someone goes to an abortion clinic. Or an NRA meeting. Or an AA meeting." Wessler goes on to say, "When police are using it to track the location of a phone, it inherently collects information not just about that phone, but about every phone in the area. It looks a whole lot like a dragnet search."

The ACLU compiled the status of cell phone location laws by state, showing how lax state legislatures have been on related individual privacy concerns. Only eight states (California, Maine, Minnesota, Montana, New Hampshire, Rhode Island, Utah, and Vermont) have laws requiring a warrant to obtain all forms of cell phone location information. Three states (Illinois, Indiana, and New Jersey) require warrants for real-time location tracking. Massachusetts requires a warrant for historical cell phone location information. Four states (Colorado, Connecticut, Delaware, and Pennsylvania) offer some limited form of cell phone location protection. Three states

(Florida, Maryland, and Virginia) currently have conflicting state and federal authorities regarding cell phone location information. In all the remaining states, either there is no binding authority around location information, such information is unprotected, or no warrant is required to obtain the information.

On June 22, 2018, a landmark cell location case, *Carpenter v. United States*, was settled by the U.S. Supreme Court, ruling that the FBI had violated Timothy Carpenter's privacy rights. The decision held that the FBI had committed a Fourth Amendment search without a probable cause–supported warrant when the government obtained 12,898 location points over a four-month period from Carpenter's wireless carriers. That cell location data led to his conviction based on his location records: it confirmed that Carpenter had been "right where the... robbery

was at the exact time of the robbery."

This 5-4 ruling is regarded as a narrow victory for privacy rights advocates, as it is limited to locations obtained through historical cell location records under provisions of the Stored Communications Act. The Supreme Court ruling states: "This decision is narrow. It does not express a view on matters not before the Court... or call into question conventional surveillance techniques and tools, such as security cameras; does not address other business records that might incidentally reveal location information; and does not consider other collection techniques involving foreign affairs or national security." The ruling does not affect the legality of cell tower simulators, which provide live recordings of cell phone locations to law enforcement—only the gathering of such locations from historical business records of wireless providers. ●

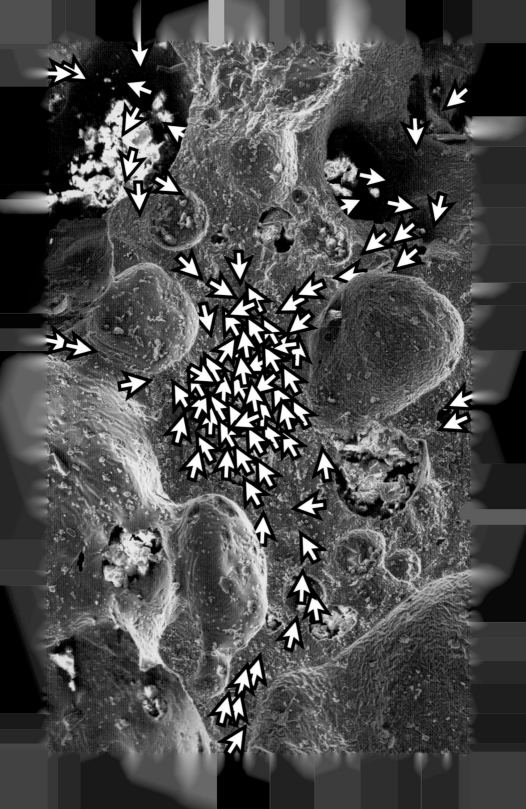

The Media Virus, My Problem Child

Douglas Rushkoff

When I published my book *Media Virus* in 1994, the most common question I got from readers was whether this new phenomenon of media viruses—through which ideas could circumvent any top-down control and spread purely based on social contagion—was "a good thing or a bad thing." I tried not to sound too enthusiastic at the time lest I betray my countercultural roots, but deep down I thought that viral media was going to change things for the better: the informational tyranny of William Randolph Hearst and Rupert Murdoch would be broken by new armies of media activists armed with photocopiers, fax machines, video cameras, cable TV, personal computers, and email messages. I saw ideas spreading as never before. They were moving laterally between people, and mutating along the way as if in a game of telephone. When one caught on, it spread like wildfire, kudzu vines... or a virus.

To me, a media virus was a sort of truth serum. It would only activate people and spread through society if it was triggering and releasing some repressed cultural agenda. Even if a viral infection made us sick, eventually it would force us to address the issues we were ignoring and begin the conversations we needed to have.

A black man getting beaten by white cops in Los Angeles happens to be captured on videotape, and the footage makes it to the cable news before morning. Smaller and tabloid media outlets do not hesitate to broadcast it and, once they do, everyone else must as well. The original "media virus" is launched and is so contagious that it leads eventually to full-scale uprising in a dozen American cities.

The inability of mainstream media's gatekeepers to control our conversation about racial injustice, inner-city police brutality, and a biased justice system was worthy of our attention. The handful of corporate conglomerates that owned almost all of the media were no longer in control of what we saw.

At the time, I was pretty sanguine about the shift from traditional media, such as newspapers and television, to interactive media like faxes, camcorders, and the internet. It felt like no one recognized the profound changes underway. My very first book on the emerging digital landscape, *Cyberia*, had been canceled because the publisher thought the internet would be "over" by 1992, when the book was scheduled for release. It seemed as if the traditional gatekeepers of media were not simply ignorant of the tidal shift underway, but actively trying to prevent it.

As many of us hoped and dreamed, interactive media wreaked havoc on those attempting to package our truth from above. It undermined the credibility of traditional news media outlets and the corporations behind them. Cable channels such as CNN were willing to put a microphone in front of dictators who were effectively censored by the broadcast network news. Public access channels created forums for taboo issues from AIDS to Iran-Contra. Scandals from Camillagate (a leaked sexy phone call between Prince Charles and Camilla Parker Bowles) to Gennifer Flowers (one of then-candidate Bill Clinton's first revealed mistresses) only made their way to the mainstream because they were first spread by the growing power of viral and bottom-feeder outlets. Even one-way TV like *Jerry Springer*, *COPS*, and MTV's *Real World* allowed for more spontaneous, uncensored content to rise from the fertile cesspools of our cultural subconscious. The traditional news media–as well as the political and corporate institutions it supported–were under threat.

But interactive media did all this bottom-up destruction without ever coalescing into something new. It was great at eroding our trust in institutions but didn't do much to engender trust between us people. We were left with a media environment where sensationalist images, facts, rumors, and ideas compete against one another for attention, agitating everyone into a state of confusion and rage, but bringing us no closer to anything resembling truth. I borrowed from evolutionary biologist Richard Dawkins and labeled this new sort of contagious idea a "meme." This was way before those funny captioned pictures people now spread on social media. Memes are really just ideas, understood from the perspective of their virality, or ability to get replicated. A meme is to an idea as a gene is to a trait, or code is to an application.

Media viruses were supposed to be anathema to traditional propaganda. Instead, in this era of Facebook, Cambridge Analytica, and cyberwarfare, they're used and recognized more and more as the latest weapons in the traditional propagandist's arsenal. A tool for countercultural expression has become the leading form of psyops.

This isn't what I was thinking. The media virus is my problem child–a powerful way of spreading necessary ideas but also, simultaneously, an evasion of our higher faculties, leading us to act out automatically, impulsively, and destructively. In other words, it's not the particular idea or meme that is so destructive–it's the viral methodology itself. The counterculture used humor, irony, or just the novelty of new media technologies to get attention for its ideas and to promote their replication. That's how phenomena such as smart drugs, flash mobs, chaos magic, and even Occupy were disseminated. The government and corporate propagandists behind today's viral warfare, on the other hand, are targeting psychological triggers, deeply embedded fears, racist bigotry, and other repressed anxiety to get attention and traction. Showing immigrant children behind bars in Texas holding centers may appear like bad PR, but it makes sense when it's understood as an intentionally viral reframing of refugees as animals or criminals. This is what we've been subjected to since at least the 2016 election–and the collateral damage is enormous. Media viruses themselves went viral, attacking not only our society's corrupt institutions but also our social bonds, our sense of common purpose, our trust in one another, and the very mechanisms we need to resist their influence.

So, whether or not memes are weaponized–no matter the intentions of their engineers–they still contribute to an overall environment of competitive propaganda and kneejerk reactions that discourages the listening, analysis, and consideration required for productive dialogue. Even viruses as legitimately formulated as #blacklivesmatter or #metoo can't solve the problems they mean to address–not alone, anyway. They can raise awareness and, at best, provoke a conversation that needs to take place. But they can do this only if they haven't created too much damage in their own right for that more thoughtful and sympathetic exchange of ideas to take place.

I feel bad about this, I truly do. But I also feel that if people had listened to what I was trying to say about viral media back in the early '90s–that it's not just a new form of content but a new way of stimulating a cultural reaction—we'd be in a much better position now to mitigate their effects on us. Most people are still looking at fixing the problem of viral media by regulating industry, writing new algorithms, or changing technology in some way, as if this were an engineering fix. That's mistaken. The only real cure for viral attack is to bolster our own immune response.

The way we choose to use media always comes down to the way we choose to use people.

Since at least Biblical times, media was understood as a form of crowd control. The notable exception was during the Enlightenment, when widespread literacy was accepted as necessary to a functioning democracy. If people were going to be allowed to vote, they'd have to be rendered informed and intelligent enough to do so.

But by the twentieth century, this progressive understanding of media literacy had been subsumed by more pressing agendas. Woodrow Wilson had run for president on a peace platform, but after he took office he decided that the United States had no choice but to enter World War I. So he hired Walter Lippmann, the father of public relations—yes, the actual guy who came up with the term "public relations"—to manufacture public consent for the war.

There was a lot of consternation about treating the masses as something to be engineered. Lippmann wrote a book called *Public Opinion*, in which he argued that people were really just too uninformed and uneducated to make the best decisions for themselves. Not even politicians could be trusted with complicated policy. Instead, a "council of experts" should be employed to figure things out and explain the necessary actions to elected officials who, in turn, would hire public relations specialists to convince the public of the chosen action.

Lippmann didn't mean this cynically. He saw himself as a progressive, using experts and government and PR men to get the common folk to express their better natures—with a little help from manipulative media. His predecessor, legendary spin doctor Edward Bernays, took this logic a bit further: people are just too stupid to know what's good for them. The future belongs to those who can control the behavior of the masses. Not only that, but this was in itself a noble pursuit, because people can't be trusted to act in their own self-interest anyway. Bernays and other advocates of appropriately applied propaganda used the Nazis as a case study of what happens without drastic measures to control the collective psyche of one's population. (Though most of us would argue it's proof of the opposite.)

Government and corporate elites feared that, unchecked, American laborers and traumatized WWII veterans could easily become the irrational mob described in Gustave Le Bon's famous 1895 book *The Crowd*. Controlling their behavior through media seemed humane compared with overt repression. Besides, media could also stoke the consumer behavior required to support American industry, create jobs, and keep the economy growing. Whether it's government pushing a policy or advertisers pushing a product, our top-down media from radio and TV to magazines and movies has been used to pump us with ideas and aspirations to make us act in certain ways.

At first glance, the horizontal landscape of interactive and social media seemed to promise more lateral communication between peers and less propaganda from above. This is what we were all celebrating in those early, heady days of the internet. The elites who owned traditional media outlets would no longer be able to serve as gatekeepers for what the masses read and watched. Anyone with a camcorder or email account would be able to get a message out. And if it was compelling enough, it would be replicated and spread to millions—without the willing cooperation of traditional media organizations.

The term "media virus" meant to convey this new way ideas could spread in a world with more interactive communications. It was like a computer virus, except instead of infecting computer networks it infected human networks. The earliest examples included people like O. J. Simpson, Madonna, Michael Jackson, or Woody Allen; ideas like smart drugs, slackers, or fractals; and things like Pogs, Beanie Babies, or emoticons. They all had spreadability and mutability. While they were not occurring on strictly interactive platforms, they reflected the morphing, lateral, peer-to-peer, bottom-up qualities of the emerging internet culture. Michael Jackson's video for "Black or White" showed people morphing (a new computer effect at the time) into other people of different races and genders. Madonna co-opted the gestural language of an underground gay culture for her "Vogue" video and disseminated it through the mainstream via MTV. The infamous slow-motion police chase of O. J. Simpson's white Bronco, as well as cable TV's gavel-to-gavel coverage of his trial, initiated the always-on quality of today's twenty-four-hour news cycle. This was the new cultural soup in which viruses would compete for attention and dominance. When our culture became immune to one version of Madonna or Michael Jackson, a new one would spawn. But–as the advertisers who quickly jumped on the idea of viral media didn't realize–it goes much deeper than this.

For a real, biological virus to infect us, it must have a novel, never-before-seen protein shell that lets it travel through our bloodstream unrecognized. (If our body identifies the virus, it sends antibodies to attack it.) The undetected virus then latches onto a cell in the host organism and injects its genetic code inside. The code is, basically, genetic material that wants to get reproduced. So it works its way to the cell's nucleus and seeks to interpolate itself into the cell's DNA. It looks for weak spots, then nests there. The next time the cell reproduces, it replicates the virus's code along with its own.

Then the person carrying the virus begins spreading it to others. If the next person's immune system doesn't recognize the protein shell, then they get infected, too. The virus continues to replicate and spread until, at last, our bodies learn to reject its code. From then on, our bodies will recognize and attack this protein shell–even if it comes back months or years later. Immunity.

A media virus works the same way. It has a novel, unrecognizable shell–but that shell is made of media, not protein. The virus must be packaged sensationally, as part of a unique, rule-breaking use of media that we can't help but spread. A camcorder tape captures police brutality. A former football star gets caught in a slow-motion chase with police on live national TV. A voice mail message reveals an actor's abusive relationship or an affair between royals. A TV star posts social media updates on his mental breakdown. An underwear commercial veers too close to child pornography. A rock album is rumored to contain hidden Satanic messages. A political candidate's wireless microphone records him making sexist remarks about a female colleague. A woman livestreams her husband dying of gunshot wounds. A congressman transmits smartphone pictures of his genitals to a minor. A Shakespeare play is reinterpreted as a presidential assassination. A president threatens a nuclear attack in a public, 140-character message typed with his thumbs.

In each case, the story's initial proliferation has more to do with the medium than with the message. The viral shell is not just a media phenomenon, but a way of grabbing attention and paralyzing a person's critical faculties. What the...? Did a white man just morph into a black woman? Is that a tweet of a congressman's erect penis? Is that really the Prince of England's answering machine message? What is that—police bodycam footage? That moment of confusion creates the time and space for infection. This "confusion technique" was first described by psychologist Milton Erickson as the primary tool for hypnotic induction. A popular version, called handshake induction, involves a hypnotist interrupting a known behavior or sequence—like shaking someone's hand or tying a shoe—and inserting something new. The break in continuity, the departure from the known and practiced script, creates a vulnerability.

Once it has been launched, once that confusion creates a pause, the virus replicates only if its code can successfully challenge our own. That's why the ideas inside the virus—the memes—do matter. They must interpolate into our own confused cultural code, exploiting the issues we haven't adequately addressed as a society, such as racial tension, gender roles, economic inequality, nationalism, or sexual norms. A fatal car crash on the side of the highway attracts our attention because of the spectacle, but worms its way into our psyche because of our own conflicted relationship with operating such dangerous machinery ourselves, or because of the way it disrupts our ongoing, active denial of our own mortality.

Likewise, a contagious media virus attracts mass attention for its spectacular upending of TV or the internet, but then penetrates the cultural psyche by challenging collectively unresolved or repressed anxieties. Surveillance video of a police van running over a black suspect recalls America's shamefully unacknowledged history of slavery and ongoing racism. The social media feed of a neo-Nazi bot in Norway stimulates simmering resentment of the European Union's dissolution of national identities. Sexual harassment via social media by a sitting president provokes the animus of a population still resentful of women in the workplace.

When I first used the expression "media virus," I thought I was describing a new sort of total transparency; media would finally tell the stories that our controllers didn't want us to hear. If a cultural issue is truly repressed or unresolved, a media virus invoking that issue can nest and replicate.

The perplexing thing–the part I didn't fully understand until now–is that it doesn't matter what side of an issue people are on for them to be infected by the meme and provoked to replicate it. "Look what this person said!" is reason enough to spread it. In the contentious social media surrounding elections, the most racist and sexist memes are reposted less by their advocates than by their outraged opponents. That's because memes do not compete for dominance by appealing to our intellect, our compassion, or anything to do with our humanity. The media space is too crowded for thoughtful, time-consuming appeals. When operating on platforms oversaturated with ads, memes, messages, spam, and more, memes need to provoke an immediate and visceral response to get noticed. "The Clintons are running an occult child sex ring in the basement of a pizzeria." In a race to the bottom of the brain stem, viruses compete to trigger our most automatic impulses.

Well-meaning and pro-social counterculture groups from the Situationists to Adbusters and Greenpeace have attempted to spread their messages through the equivalents of viral media. They cut and paste text and images to subvert the original meanings of advertisements, or the intentions of corporate logos. It is a form of media aikido, leveraging the tremendous weight and power of an institution against itself with a single clever twist. With the advent of a new, highly interactive media landscape, internet viruses seemed like a great way to get people talking about the unresolved issues that needed to be discussed in the light of day. After all, this logic goes, if the meme provokes a response, then it's something that has to be brought up to the surface.

But we can't engineer a society through memetics the way a biologist might hope to engineer an organism through genetics. It's ineffective in the long run, and–beyond that–unethical. It bypasses our higher faculties, our reasoning, and our collective authority.

The danger with viruses is that they succeed by bypassing the neocortex–the thinking part of our brain–and go straight to the more primal reptile beneath. The meme for scientifically proven climate change, for example, doesn't provoke the same intensity of cultural response as the meme for "elite conspiracy!"

Logic and truth have nothing to do with it. Memes work by provoking fight-or-flight reactions. And those sorts of responses are highly individualistic. They're not pro-social; they're antisocial. They're not pro-cultural; at their best they are countercultural. They can galvanize a particular group of people, especially one that feels under assault. If the group is genuinely vulnerable–such as #BlackLivesMatter, #MeToo, or #ArabSpring–then this solidarity, though usually emotional and oppositional, is still beneficial to the group's identity and cohesion. But the very same memetic provocations work, perhaps even better, to galvanize groups on false pretenses. As long as the deep fear, rage, or panic is activated, it doesn't have to be based in reality. Indeed, fact-based rhetoric only gets in the way of the hyperbolic claims, emotional hot buttons, and mythic claims that rile people up: blood and soil, black men will hurt you, foreigners are dangerous, Lock Her Up. The less encumbered by facts or sense, the more directly a meme can focus on psychological triggers from sexism to xenophobia.

So, for example, a viral assault is not likely to persuade a bankrupted town of unemployed coal workers to adopt strategies of mutual aid. It could, on the other hand, help push the disenfranchised toward more paranoid styles of self-preservation. With notable exceptions–such as the Twitter messages of support during the failed Iranian protests, or those between Ariana Grande fans after her concert was bombed in Manchester–memetic campaigns do not usually speak to the part of the brain that understands the benefits of tolerance, social connection, or appreciation of difference. They're speaking to the reptile that only understands predator-or-prey, fight-or-flight, and kill-or-be-killed. Even these positive exceptions were in response to something as shocking and horrible as any meme.

The bottom-up viral techniques of guerrilla media activists are now in the hands of the world's wealthiest top-down corporations, politicians, propagandists, and everything in between. To them, viral media is no longer about breaking through propaganda and unearthing the truth about social inequality or environmental threats. It's simply about generating a response by any means necessary, even if that response is automatic, unthinking, and brutish. Like a military using chemical weapons that spread to its own troops, we are using a weapon that we do not understand, and at our own collective peril.

We have to remember that the concept of memetics was first popularized not by a cultural anthropologist, poet, or media theorist but by a particularly materialist evolutionary biologist in the 1970s. A strident atheist, Dawkins meant to show how human culture evolves by the same set of rules as any other biological system: competition, mutation, and more competition. Nothing special going on here.

It turns out there *is* something special going on here, and that there are a few things missing from this explanation. A meme is a great corollary to a gene, for sure, but neither genes nor memes determine everything about an organism or a culture.

DNA is not a static blueprint but acts differently in different situations. It matters which genes we have, but it matters even more how those genes express themselves. That's entirely dependent on the environment, or the protein soup in which those genes are swimming. It's why a locust can be like a tame grasshopper or, in the right conditions, transform into a gregarious, swarming creature. That's not a sudden mutation within a single lifetime; it is a shift in gene expression that changes the whole organism.

Genes are not solo actors with entirely predetermined code. They are not selfishly seeking their own replication at all costs. Newer science shows they are almost social in nature, adapting and expressing themselves differently in different environments. Organisms get information from the environment and from one another for how to change. The conditions, the culture, and its connectivity matter as much as the initial code.

Similarly, if we truly want to understand cultural contagion, we must place equal importance on the viral shell around memes and on the ideological soup in which those memes attempt to spread. Early memeticists saw memes as competing against one another, but that's not quite right. Memes are all attempting to self-replicate by exploiting inconsistencies or weaknesses in our cultural code. They are not attacking one another; they are attacking us humans.

Advertising agencies loved that earlier explanation, because it meant all they had to do was work on crafting the most contagious meme for it to "go viral." But that's not how it actually works, and why most of those campaigns failed miserably. A famous 2005 web video ad of Paris Hilton washing a car in a bathing suit may have reached a wide audience, but it did little long-term good for the brand of hamburgers it was supposed to be spreading. Neither did dozens of copies of dancing babies, cute cats, rapping cereal characters, or opportunities to vote for the color of future soft drinks—all meant to go viral. They are cute enough to look at or even pass on to a friend, but they don't have any embedded content—nothing to challenge our existing cultural code. Cats are cute. Teenage boys like to look at girls in bikinis. There's nothing under the surface to be unleashed by any of this. On the other hand, the Calvin Klein underwear ads made to look like child-porn film shoots (which were subsequently pulled) succeeded in generating millions of dollars' worth of secondary media, and in reestablishing the brand's rebellious image. (As I later learned, the creatives responsible for those ads had based them on the principles of my book. #mixedfeelings) The point is, a meme can go viral only if it is unleashing or leveraging a repressed cultural agenda or taboo. The potential has to be there already.

The Trump viral shell was his reality-show persona and its unique migration to real-world politics. But the memes within the Trump virus replicated–at least in part–because there was already a widespread, though still partially pent-up, white nationalist rage in America.

Human societies must come to recognize the importance of developing a healthy cultural immune response to an onslaught of hostile memes. The technologies through which they are being transmitted are changing so rapidly that it would be impossible to recognize their new forms–their shells–in advance. We must instead build our collective immune system by strengthening our organic coherence–our resistance to socially destructive memes.

This is particularly difficult when the enemies of democracy and their unwitting allies (the communications directors of political campaigns) are busy upscaling memetic warfare with each of social media's latest tricks, from data mining to predictive algorithms. In addition to artificially amplifying the "scale" of memes that may not have gained any organic traction on their own, these algorithms and bots are designed to engage with us individually, disconnect us from one another, neutralize our defense mechanisms, and program our behaviors as if we were computers. Television advertisers may have normalized the idea that consumers can be experimented on like lab rats, but social media takes it to an entirely new level.

At least advertising through TV happens in public. TV ads are expensive, proving that there is a big company behind the product willing to invest in its success, and TV stations censor ads they find offensive. Social media manipulates us individually, one smartphone at a time. Posts may cost pennies or nothing at all, and they're sold and placed by bots with no regard for their content. When media is programmed to atomize us, and the messaging is engineered to provoke our most competitive, reptilian sensibilities, it's much harder to muster a collective defense.

The powers working to disrupt democratic process through memetic warfare understand this well. Contrary to popular accounts, they invest in propaganda from all sides of the political spectrum. The particular memes they propagate through social media are less important than the reactions they hope to provoke. The Russians sent messages to Bernie Sanders supporters meant to stoke their outrage at the Democratic party's favoritism toward Hillary and discourage them from voting in the general election. After school shootings, Russian and other bots begin pumping out extremist messages on both sides of the gun debate. Fake news spread about the Parkland shooter's supposed terrorist ties, as well as falsified links to the anti-fascist group Antifa. They're intended not to promote meaningful debate, but to exploit an opportunity to incite fear, disable rational thinking, and provoke ideological clashes. The shootings are an opportunity to undermine civil discourse and social cohesion.

Memetic warfare, regardless of the content, discourages cooperation, consensus, or empathy. The reptile brain it triggers doesn't engage in those pro-social behaviors. Instead, in an environment of hostile memes and isolated by social media, human beings become more entrenched in their positions and driven by a fear for their personal survival. Worst of all, since these platforms appear so interactive and democratic, we experience this degradation of our social processes as a form of personal empowerment. For some, to step out of the corner and be truly social starts to feel like a restraint—as if yoked by political correctness, or forced into showing compromising tolerance of those whose very existence "weakens our stock." Progressives, likewise, find solace in their own online echo chambers, and use the worst examples of far-right troll behavior to justify their intolerance of anyone who identifies with red-state values. Anyone who uses the hashtag is one of us; those who don't, well, they're the enemy.

Traditional media, like television, urged us to see the world as one big blue marble. Ronald Reagan could go on television, stand in front of the Berlin Wall, and demand that Mr. Gorbachev "tear down this wall!" In the divisive world of memetic digital media, Donald Trump can tweet his demand that we *build* a wall to protect us from Mexico. Virality encourages less connection, intimacy, and cross-contamination. Progressives are sensitized, through memetics, to every misunderstanding of their racial, gender, or cultural identity. Trumpists, meanwhile, are pushed toward a counterphobic urge to call all Muslims terrorists or all Mexicans gang members. Memetics helps them see institutions from the FBI to government itself as a vast conspiracy against their leader.

This may not have been the intent of social media, or any of the communications technologies that came before it. The internet doesn't have to be used against a person's critical faculties any more than language has to be used to lie or numbers to tally enslaved people. But each extension of our social reality into a new medium requires that we make a conscious effort to bring our humanity along with us.

A few years ago I had dinner with a former U.S. secretary of state. We were debating America's ability to conduct itself democratically. Were Lippmann and Bernays right? Could the masses simply not be trusted? This was long before the era of Trump, mind you. But Fox News was already in full swing, and the antics of public relations engineers had reached new heights.

We were talking about the first Gulf War, and how a PR firm called Hill & Knowlton had not only made up a story about Iraqi soldiers pulling premature babies from their incubators and leaving them to die, but also gotten a diplomat's daughter to testify before Congress, pretending that she had witnessed the atrocities. The video of her tearful testimony about the babies being left "to die on the cold floor" of the hospital went viral, and America went to war. So much for my theory of media viruses having a positive effect on public debate.

The old statesman finally turned to me and grinned. "So, Rushkoff, do you now accept the fact–beyond any shadow of a doubt–that democracy has been proven a failed experiment?" I was shocked that he even remembered my name. Something about his power and reputation silenced me, and the conversation went on to something else.

But, no, I'm not ready to concede that democracy was a failed experiment, or that human beings have been proven incapable of governing themselves.

We may not be doing so well at the moment, but we mustn't surrender to the notion that we or our political adversaries are constitutionally incapable of engaging in meaningful dialogue or making informed decisions in our mutual best interest–especially when that argument is being made by the very people who have taken it upon themselves to manipulate our thoughts, feelings, and actions by any means necessary.

After all, political campaigns have always relied on values, visions, narratives, and ideologies to win votes. Whether virtuous or cynical, this effort comes down to propaganda: the leverage of social and psychological biases to promote a particular point of view. Any technique, from a Hearst banner headline to a Cambridge Analytica-engineered virus, seeks to reach down into our brain stem and trigger us to behave in reactionary, robotic ways. And the collateral damage of these assaults is the same. Memetics is just the latest tool for engineering the same old compliance.

We have three main choices for fighting back.

The first is to attack bad memes with good ones. Tit for tat. While such an approach may be appropriate in a crisis, the problem is that it increases the amount of weaponized memetics in play at any particular time. The enemy memes may be weakened, but so, too, is the community of humans under attack.

The second choice is to try to insulate people from dangerous viruses–the same way a person might wear a surgical mask in an airport. In the media landscape, that means adding new filters, algorithms, and digital countermeasures to the latest and greatest innovations of the social-media companies and the market research firms paying them for their data. So if we know that Russian propagandists are paying Facebook to deliver provocative false stories to our news feeds, we install a filter that weeds out unsourced stories, or an algorithm that uses machine learning to identify common word choices in fake-news posts.

But an arms race of this sort just pits one side's black-box technologies against another's. It's more like today's stock market: a war between computer engineers. May the best algorithm win. The battle for our hearts and minds ends up occurring on a level far removed from that of civic discourse. Ideas and candidates don't win on their merits, but on their digital gamesmanship. That may be some nerd's idea of a win-win–kind of like Bitcoin–but it's not democracy.

A less dramatic but ultimately more powerful approach is to strengthen the cultural immune response of the society under attack. This could mean educating people about the facts around a particular issue, or bringing very controversial but memetically potent issues into the light of day. Schools can teach classes from the Courageous Conversations curriculum. Towns can use consensus-building tools like the Loomio platform to discuss and address issues that get needlessly polarized in social-media channels. Politicians can choose to articulate the real anxieties fueling the racist or xenophobic stances of their adversaries, rather than pretending such feelings simply don't exist. A society having an open, honest conversation about race, guilt, and fear of change is less vulnerable to a memetic attack invoking white supremacy than a society still afraid to have that painful conversation.

Bringing repressed issues up and out into the light of day reduces the potential difference–the voltage–between the expressed and unexpressed cultural agendas of that moment. The hostile memes will either not be able to locate confused code in which to nest, or, if they do, fail to produce a rapid acceleration of reproduction.

The downside to such strategies, of course, is the question of whose curriculum is used to educate the public about a particular issue. Town halls and other public forums are great for airing grievances, but at some point the conversation will have to turn to real history, real facts, or real science. Whose real is accepted? We end up back in the highly criticized situation envisioned by the father of public relations, Walter Lippmann: his council of experts informing government officials of the appropriate action, and an army of public relations specialists engineering public consent.

In the currently militarized sociopolitical environment, any efforts at education would be interpreted as partisan at best, and elitist and untrustworthy at worst. But this doesn't mean we should give up. It just means we may have to pull our attention from memes themselves, and examine instead the conditions in which they either grow or peter out.

The longest-term strategy to defend against memetic attack, and ultimately the most effective one, is to strengthen the social and cultural resiliency of the population under attack—whether it's an underserved rural white community susceptible to neo-Nazi memes or an African American community whose vulnerability to anti-police memes has been primed by years of stop-and-frisk abuse. Human beings have evolved complex and adaptive strategies for social cohesion. Our neurology is primed to establish rapport with other humans, to utilize reciprocal altruism, and to work toward common goals. Such social relationships require real-world, organic calibration to take effect. They can be amplified by social media, but they must be anchored in the natural world lest they become too brittle, abstract, or mutable, and easily co-opted by someone with very different goals and values.

The establishment of rapport, for example, depends on eye contact, synchronized respiration, and recognition of subtle changes in vocal timbre. In virtual spaces, these mechanisms cease to function. In fact, when human beings fail to establish "social resonance" through digital media, they tend to blame not the low fidelity of the medium, but the trustworthiness of the other party. Hear that: the inability to establish organic social bonds through digital media increases our suspicion of one another, not of the medium through which we are failing to connect.

This creates the perfect preconditions for memetic attack. The people, newscasters, friends, and experts we encounter through digital media are not trusted. Their faces don't register as faces, so we reject their honesty. The faceless bots, algorithms, images, and ideas to which we are exposed, on the other hand, are accepted at face value because they don't trigger that same cognitive dissonance. There's no face not to trust—just the fake facts and sensationalist vitriol, feeding straight down into the brain stem.

The only surefire safeguard against this state of vulnerability is to reaffirm the live, local, social, organic relationships between the people in the target population. This means challenging the value of time spent socializing on digital platforms, and giving people enough minutes of non-digitized social experiences each day to anchor live human-to-human connection as the primary form of social engagement.

People with some live experience of local politics, mutual aid, and environmental maintenance will be more resistant to the memetic constructions of the synthetic ideological landscape. They will be more likely to blame low fidelity on technology than on one another, and less likely to accept the false, antisocial premises of angry, sensationalist memes. Of course, local social cohesion doesn't always translate to tolerance of others. Loyalty to one's "hometown" already suggests favoritism to one's neighbors and a bit of suspicion about anyone from somewhere else. But if we're going to see pro-social attitudes and behaviors ever get to "scale," they must be intentionally and formally embedded in our platforms and the standards we establish for ourselves when using them. In order to do this, we absolutely must reacquaint ourselves with what it feels like to establish rapport, reach consensus with the opposition, and trust that what looks like hate is likely coming from a place of fear.

The less alienated the members of a population are from one another, the harder it is to turn them against one another. We start to trust our senses again, as well as our relationships, our critical faculties, and the notion of truth itself. ●

Boeing engineers prepare a GPS satellite for launch.

The Right to Experiment

Bruce Schneier

n my book *Data and Goliath*, I write about the value of privacy. I talk about how it is essential for political liberty and justice, and for commercial fairness and equality. I talk about how it increases personal freedom and individual autonomy, and how the lack of it makes us all less secure. But this is probably the most important argument as to why society as a whole must protect privacy: it allows society to progress.

We know that surveillance has a chilling effect on freedom. People change their behavior when they live their lives under surveillance. They are less likely to speak freely and act individually. They self-censor. They become conformist. This is obviously true for government surveillance, but is true for corporate surveillance as well. We simply aren't as willing to be our individual selves when others are watching.

Let's take an example: hearing that parents and children are being separated as they cross the U.S. border, you want to learn more. You visit the website of an international immigrants' rights group, a fact that is available to the government through mass internet surveillance. You sign up for the group's mailing list, another fact that is potentially available to the government. The group then calls or emails to invite you to a local meeting. Same. Your license plates can be collected as you drive to the meeting; your face can be scanned and identified as you walk into and out of the meeting. If instead of visiting the website you visit the group's Facebook page, Facebook knows that you did and that feeds into its profile of you, available to advertisers and political activists alike. Ditto if you LIKE their page, share a link with your friends, or just post about the issue.

Maybe you are an immigrant yourself, documented or not. Or maybe some of your family is. Or maybe you have friends or coworkers who are. How likely are you to get involved if you know that your interest and concern can be gathered and used by government and corporate actors? What if the issue you are interested in is pro- or anti-gun control, anti-police violence or in support of the police? Does that make a difference?

Maybe the issue doesn't matter, and you would never be afraid to be identified and tracked based on your political or social interests. But even if you are so fearless, you probably know someone who has more to lose, and thus more to fear, from their personal, sexual, or political beliefs being exposed.

This isn't just hypothetical. In the months and years after the 9/11 terrorist attacks, many of us censored what we spoke about on social media or what we searched on the internet. We know from a 2013 PEN study that writers in the United States self-censored their browsing habits out of fear the government was watching. And this isn't exclusively an American event; internet self-censorship is prevalent across the globe, China being a prime example.

Ultimately, this fear stagnates society in two ways. The first is that the presence of surveillance means society cannot experiment with new things without fear of reprisal, and that means those experiments–if found to be inoffensive or even essential to society–cannot slowly become commonplace, moral, and then legal. If surveillance nips that process in the bud, change never happens. All social progress–from ending slavery to fighting for women's rights–began as ideas that were, quite literally, dangerous to assert. Yet without the ability to safely develop, discuss, and eventually act on those assertions, our society would not have been able to further its democratic values in the way that it has.

Consider the decades-long fight for gay rights around the world. Within our lifetimes we have made enormous strides to combat homophobia and increase acceptance of queer folks' right to marry. Queer relationships slowly progressed from being viewed as immoral and illegal, to being viewed as somewhat moral and tolerated, to finally being accepted as moral and legal. In the end it was the public nature of those activities that eventually slayed the bigoted beast, but the ability to act in private was essential in the beginning for the early experimentation, community building, and organizing.

Marijuana legalization is going through the same process: it's currently sitting between somewhat moral, and–depending on the state or country in question–tolerated and legal. But, again, for this to have happened, someone decades ago had to try pot and realize that it wasn't really harmful, either to themselves or to those around them. Then it had to become a counterculture, and finally a social and political movement. If pervasive surveillance meant that those early pot smokers would have been arrested for doing something illegal, the movement would have been squashed before inception. Of course the story is more complicated than that, but the ability for members of society to privately smoke weed was essential for putting it on the path to legalization.

We don't yet know which subversive ideas and illegal acts of today will become political causes and positive social change tomorrow, but they're around. And they require privacy to germinate. Take away that privacy, and we'll have a much harder time breaking down our inherited moral assumptions.

The second way surveillance hurts our democratic values is that it encourages society to make more things illegal. Consider the things you do–the different things each of us does–that portions of society find immoral. Not just recreational drugs and gay sex, but gambling, dancing, public displays of affection. All of us do things that are deemed immoral by some groups, but are not illegal because they don't harm anyone. But it's important that these things can be done out of the disapproving gaze of those who would otherwise rally against such practices.

If there is no privacy, there will be pressure to change. Some people will recognize that their morality isn't necessarily the morality of everyone–and that that's okay. But others will start demanding legislative change, or using less legal and more violent means, to force others to match their idea of morality.

It's easy to imagine the more conservative (in the small-*c* sense, not in the sense of the named political party) among us getting enough power to make illegal what they would otherwise be forced to witness. In this way, privacy helps protect the rights of the minority from the tyranny of the majority.

This is how we got Prohibition in the 1920s, and if we had had today's surveillance capabilities in the 1920s it would have been far more effectively enforced. Recipes for making your own spirits would have been much harder to distribute. Speakeasies would have been impossible to keep secret. The criminal trade in illegal alcohol would also have been more effectively suppressed. There would have been less discussion about the harms of Prohibition, less "what if we didn't..." thinking. Political organizing might have been difficult. In that world, the law might have stuck to this day.

Effects of Government Surveillance On Writers' Self-Censorship

28%
◀ Writers who curtailed or avoided social-media activities

24%
◀ Writers who deliberately avoided certain topics in phone or email conversations

16%
◀ Writers who refrained from making internet searches or visiting websites on topics that may be considered suspicious

16%
◀ Writers who avoided writing or speaking on a particular topic

Source: PEN America

▲ Four months after the 2013 revelations about NSA spying, writers reported changing their practices because they thought the government may be monitoring them.

China serves as a cautionary tale. The country has long been a world leader in the ubiquitous surveillance of its citizens, with the goal not of crime prevention but of social control. They are about to further enhance their system, giving every citizen a "social credit" rating. The details are yet unclear, but the general concept is that people will be rated based on their activities, both online and off. Their political comments, their friends and associates, and everything else will be assessed and scored. Those who are conforming, obedient, and apolitical will be given high scores. People without those scores will be denied privileges like access to certain schools and foreign travel. If the program is half as far-reaching as early reports indicate, the subsequent pressure to conform will be enormous. This social surveillance system is precisely the sort of surveillance designed to maintain the status quo.

For social norms to change, people need to deviate from these inherited norms. People need the space to try alternate ways of living without risking arrest or social ostracization. People need to be able to read critiques of those norms without anyone's knowledge, discuss them without their opinions being recorded, and write about their experiences without their names attached to their words. People need to be able to do things that others find distasteful, or even immoral. The minority needs protection from the tyranny of the majority.

Privacy makes all of this possible. Privacy encourages social progress by giving the few room to experiment free from the watchful eye of the many. Even if you are not personally chilled by ubiquitous surveillance, the society you live in is, and the personal costs are unequivocal. ●

▷

A COMPENDIUM OF LAW ENFORCEMENT SURVEILLANCE TOOLS

By Edward F. Loomis

○ Facial Recognition Systems

○ GPS Trackers

○ License Plate Readers

○ Drones

○ Body Cameras

○ Cell Tower Simulators

● **Parallel Construction**

Parallel construction isn't so much a technology as it is a method used by law enforcement to shield evidence obtained illegally through technologies such as those described in this issue from the courts and defense attorneys. Human Rights Watch defines parallel construction as a "deliberate effort by U.S. Government bodies, as part of a criminal investigation or prosecution, to conceal the true origins of evidence by creating an alternative explanation for how the authorities discovered it." In employing this scheme, police violate a defendant's due process right to a fair trial and preclude the privacy protections of the "exclusionary rule," which says that evidence obtained in violation of a defendant's constitutional rights may not be entered into a court of law.

▷ **A Compendium of Law Enforcement Surveillance Tools**

○ Facial Recognition Systems
○ GPS Trackers
○ License Plate Readers
○ Drones
○ Body Cameras
○ Cell Tower Simulators
● **Parallel Construction**

MCS54 **0314**

Concerns over the use of parallel construction first appeared in a Reuters article published in August 2013 referencing a DEA document used to train agents for its Special Operations Division (SOD). The document reveals that SOD agents were instructed to use drug-related communications of U.S. citizens incidentally intercepted during counterterrorism operations to initiate new drug investigations, and when doing so to disguise the originating evidence from reports, affidavits, attorneys, and court documents so as not to reveal their source—thus preventing defense attorneys from mounting objections to privacy invasion.

Concerns over parallel construction are heightened when newer technologies with inherent inaccuracies (*e.g.*, facial recognition systems) are used to initiate an investigation. If the technology is hidden from the judge, a determination cannot be made as to whether the technology violated a defendant's rights.

In cases researched by Human Rights Watch in 2016 to 2017 involving parallel construction, pretextual traffic stops were often involved. In these instances, local law enforcement tails a suspect and pulls them over for a minor violation, like failing to use a turn signal. While the stop may appear random, the police are often working on a tip received from a federal agency such as the DEA.

Revealed in a Department of Defense Inspector General 2009 publication, *Report on the President's Surveillance Program*, an FBI team was instructed in 2003 to share information from a classified NSA intelligence program with Bureau colleagues "without disclosing that the NSA was the source of the information or how the NSA acquired the information." Cases in which evidence obtained by the controversial FISA Section 702 provisions were found to have been hidden from defendants in several successful federal prosecutions (*e.g.*, *United States v. Mohamud*, case no. 3:10-cr-00475 (D. Or.); *United States v. Hasbajrami*, case no. 1:11-cr-00623 (E.D. NY); and *United States v. Kurbanov*, case no. 1:13-cr-00120 (D. Id.)).

The co-director of the Brennan Center's Liberty and National Security Program, Faiza Patel, stated, "The failure to provide notice not only prevents defendants from challenging surveillance programs in court, but also stymies the public's interest in understanding how and when its vast authorities are used."

A massive telephone-call database known as Hemisphere, containing call records dating back to 1987, is documented as an evidence source concealed by the police via parallel construction. Law enforcement officials access the database by simply submitting an administrative subpoena for its data. The data is obtained through a cooperative arrangement with AT&T, which owns an estimated 75 percent of all landline switches in the United States and a large portion of the wireless infrastructure, from which the call records are obtained. Hemisphere's database continues to grow by a reported four billion new call records daily from all carriers whose calls transit the AT&T switches. According to a sensitive LAPD slide show on Hemisphere obtained by the *New York Times* in 2013, law enforcement requesters of Hemisphere data "are instructed to never refer to Hemisphere in any official document. If there is no alternative to referencing a Hemisphere request, then the results should be referenced as information obtained from an AT&T subpoena."

Once police receive results from their Hemisphere query, they send a new subpoena directly to the suspect's phone provider to obtain records for the new phone numbers identified by Hemisphere. The police then reference only the second request for records in their reports, warrant affidavits, and court testimony.

Although the Justice Department initiated the Hemisphere program as a counter-narcotics tool, the program has been used to investigate numerous other crimes, including Medicaid fraud and murder. According to Aaron Mackey, an attorney at the Electronic Frontier Foundation, Hemisphere identifies relationships and associations and builds a social network for every phone number in its database. "It's highly likely that innocent people who are doing completely innocent things are getting swept up into this database," reported Mackey. ●

Foreseeing FOIAs from the Future with Madeline Ashby

Dave Maass

Professional corporate futurists tend to cast their foresight around ten years out, give or take, depending on the client and project. Trends in technology and culture, the divergent paths they may take toward utopia, dystopia, or status-quopia, can be most easily envisioned, at least as alternative scenarios, at that decade mark.

I'm no futurist, but I am certain about one thing: a decade from now, governments will be sitting on larger stacks of data and bureaucratic paperwork than ever before. And journalists, government watchdogs, and city-hall gadflies will consequently be more curious than ever to crack open the inner workings of the machines of authority.

These thoughts swirled in my mind throughout the 2018 SXSW (South by Southwest) Interactive Festival, in part because the festival overlapped with Sunshine Week, the annual seven-day period when the transparency community celebrates (and commiserates about) open government. I found myself sequestered in a four-bedroom homeshare located a fifteen-minute rideshare away from the activities in downtown Austin.

Fitting with the theme of many of the unofficial, offsite SXSW side events, the residence was named Sci-Fi House by its sponsors, New America's Open Technology Institute and Arizona State University's Center for Science and the Imagination. The concept was to provide a space in which to bring together, under a single roof, science fiction authors, futurists, and tech policy experts to cross-pollinate between the festivities.

By day, Sci-Fi House residents would facilitate foresight events with groups like the U.S. Conference of Mayors, and by night we'd hop between opulent, corporate-sponsored parties, snacking on deconstructed hors d'oeuvres like Facebook's chicken and waffles (a waffle cone rimmed with mashed potatoes and filled with coleslaw and a single chicken finger). And then, even later, removed from the optimistic haze of SXSW, we'd have buzz-headed debates about technology.

As a futurist whose consulting gigs traverse industries, sectors, and continents, Madeline Ashby is rarely asked to look more than ten years out. However, as a science fiction author (whose recent book *Company Town* was a finalist for

○ Dave Maass   ● Foreseeing FOIAs   0319
from the Future
with Madeline Ashby

Canada Reads), Madeline often speculates much further into the unknown. And so when we met at Sci-Fi House, I had many questions about the years to come.

Top of my mind: in 2028, will the sun still shine?

And by that, as a professional muckraker, I specifically meant, will it shine upon the government? Will truth-seekers continue to expose uncomfortable facts about the powers that be? Or will public documents become even more deeply enveloped in darkness?

Generally, most government agencies in the United States have a legal duty to provide copies of records to members of the public who request them. All a person usually needs to do is file a formal letter identifying the information they're seeking. *Of course* there are plenty of loopholes for the government to withhold things like trade secrets and national security interests, and *of course* these exceptions are often abused. But despite all that, the digital age has resulted in new ways to obtain information, including large sets of raw data compiled by government agencies.

In my work as a researcher and journalist, I submit a lot of public records requests in order to understand the threats we face as technology advances. Many of these requests are issued under the federal Freedom of Information Act, although I'll often use *FOIA* more generically to describe a request under any similar state-level law. In fact, a ten-year goal of mine is to get *FOIA* accepted into an official dictionary as a noun ("I filed a FOIA"), a verb ("I FOIAed those records"), and a modifier ("Check out these sweet FOIA docs").

On our last night in Austin, after gorging on shrimp in the blue glow of the James Cameron's *Avatar*-themed launch party for the new XPrize telepresence challenge, Madeline and I retreated to Sci-Fi House. Lingering around the kitchen island as some other family's microwave clock marked our incremental progress into the future, we began conjecturing about what kind of FOIA requests we could imagine people filing in 2028.

Madeline described scientific and technological advancements we might expect a decade down the road, and we'd dream up scenarios in which some member of the public would file a FOIA request to seek information related to those developments from the government.

TAKE-HOME ROBOTS

Dear Chief FOIA Officer,

Pursuant to the Freedom of Information Act, Veterans for Equitable Mental Health Care requests the following documents:

1. All Department of Defense (DOD) reports that examine
   cases of Automated Robotics Separation Anxiety (ARSA).
2. All DOD estimates for the cost of transferring autonomous robotics equipment to servicemembers upon their
   release from service.
3. For the fiscal years 2024-2028, the number of requests
   (Form DD5414) received from servicemembers during the
   discharge process for autonomous robotics transfers and
   the corresponding number of approvals and denials.

This request includes but is not limited to unmanned aerial vehicles, bomb-sniffing drones, and Legged Squad Support
Systems (LS3), such as BigDog.

This scenario imagines a future where semi- or fully autonomous robots have become commonplace on the battlefield.

In her book *Culture and Human-Robot Interaction in Militarized Spaces: A War Story,* Dr. Julie Carpenter examines the complicated nature of human attachment to robots and finds that many soldiers see these robots as extensions of themselves. Carpenter posed these questions in an interview with *Forbes* in 2016: "In ten or twenty years, when humanlike and animal-like robots are employed in a more drone-like way from a greater distance, will a similar user self-extension or new human-robot social phenomenon cause any hesitation during human-directed tasks and affect mission outcomes? Or, will people develop an indifference to using robots as extensions of their own physicality?"

In our scenario, we imagine a nonprofit has formed to advocate for the rights of veterans who are suffering depression due to separation from their robotic assistants. Their crisis hotline has collected anecdotal evidence that more data is needed to convince Congress to fund programs that allow veterans to "adopt" their robots. The problem is that while veterans today are often able to take home military canines, robots are far more reusable, expendable, and expensive—and the Department of Defense has been reluctant to give them up.

The organization gets a tip from a source in Veterans Affairs that the DOD has failed to publicize a damning report from the agency's inspector general regarding servicemembers and their relationships with their robots. After the DOD rejects the FOIA request, the nonprofit sues and eventually frees information that reveals that top military brass had secretly sold robots on the cheap to foreign governments at lavish yacht parties where they were provided with prostitutes and envelopes full of cash.

LOOKING FORWARD TO SEE BACKWARD

```
To the National Archives and Records Administration:

Pursuant to the Freedom of Information Act, I am seeking re-
cords from 2003 related to the following issues: yellowcake,
Valerie Plame and Joe Wilson, the Air Marshal program, the
Total Information Awareness program, and "flash mobs." Under
Executive Order 13526, these records were required to under-
go mandatory declassification review by September 30, 2028.
    Should these documents remain classified, I request that
you release the classification level that was recommended
by the machine learning algorithm during its preliminary
review, prior to the official decision by the human classi-
fication officer.
```

In the future, we'll know a lot more about our past.

Federal agencies are required to complete automatic review of classified national security records when they become twenty-five years old. As the U.S. Department of Justice puts it, the goal is to provide researchers and the public with records to enhance "their knowledge of the United States' democratic institutions and history, while at the same time ensuring that information which can still cause damage to national security continues to be protected." That means records can be released by default, unless one of a few exemptions apply, such as records that would endanger a secret informant or blueprints for weapons of mass destruction.

Do the math, and you'll discover that in 2028 the public may get its first chance to review currently classified national security documents from 2003–an important year, as it marked the U.S. invasion of Iraq.

In this scenario, we imagine a reporter who cut her teeth as a young correspondent covering Vice President Dick Cheney, and has long been troubled by unanswered questions from the Operation Iraqi Freedom era. She covered the fallout, so to speak, from the infamous July 6, 2003, *New York Times* essay in which former ambassador Joe Wilson poked holes in a key claim by the Bush administration that Iraq had purchased uranium "yellowcake" from Niger. A few years later, she reported on the trial of Scooter Libby, the Cheney aide who was accused of retaliating by exposing the identity of Wilson's wife, Valerie Plame, as a CIA operative. The leaker later turned out to be Deputy Secretary of State Richard Armitage, but she was the first reporter to file when a jury nonetheless found Libby guilty of lying to federal investigators. More than a decade after she thought this story had been laid to rest, she was called back to cover it when President Trump pardoned Libby in 2018.

In 2028, this journalist is in her fifties, researching for a book about those tumultuous years and trying to leverage the Freedom of Information Act. Beyond the Plame saga, she also wants to know about the Air Marshal program and the Pentagon's short-lived mass surveillance program, Total Information Awareness, which aimed to suck up and analyze data on millions of people to purportedly identify potential terrorists. For good measure she throws in "flash mobs," since years ago she was at a bar and had a tipsy officer (pun intended) in the newly minted Department of Homeland Security blab to her about how his colleagues perceived the new craze as a public safety threat.

With so many billions of records to process, the U.S. National Archives and Records Administration (NARA) has adopted a machine learning technology to help sort through the vault and make initial recommendations about how these documents should be classified.

This technology is currently in development by the CIA and the Center for Content Understanding at the University of Texas at Austin for this exact purpose. It's called SCIM, which stands for Sensitive Content Identification and Marking. Currently it's being trained on Reagan-era emails.

Algorithms like SCIM may have an enormous impact on transparency in the future. Done right, the software could process records more quickly than any team of humans and provide a baseline review for a process that is highly objective. Done wrong, the program could justify new levels of secrecy that would simply be rubber-stamped by a human operator. It's our hope that even if a record must be withheld, our imagined reporter would at least be able to find out whether the computer had recommended that the records be buried. If SCIM issued a dissenting opinion, the reporter could use that in appealing the decision.

After SXSW, I asked Steven Aftergood, the declassification expert at the Federation of American Scientists' Project on Government Secrecy, about the legitimacy of this request. He responded, "I think it's basically fine for purposes of this piece. Of course, an actual FOIA request (now or in the future) would not (or should not) include all of these topics in a single request."

Our reporter learns that the hard way, and in 2029 files twenty separate requests, including for the remaining redacted information in the 9/11 Commission Report, records related to CIA director George Tenet's and Secretary of State Colin Powell's resignations in 2004, and of course that year's XPrize for a non-government manned spacecraft.

WHAT'S IN YOUR SOCKET?

Dear Councilmember Gary Schmitty,

Seeing as your office has continually refused to meet with me to discuss the impending rate increase for utility users such as myself, I have no choice but to file this public records request to prove the problem is not us—your constituents—but you. I hereby demand that you provide all data sets associated with energy consumption by each and every member of your staff, broken down by office space, time, date, and device. I'm willing to bet that you're more of an energy hog than we are. Oink oink.

When you're a public servant, your professional life is an open book. Government watchdogs file for salaries and bonuses, calendars, spending requests, security detail expenses, and flight itineraries to uncover malfeasance, with great effect. Already during the Trump administration, public records have cost top officials their jobs, including Health Secretary Tom Price and Veteran Affairs Secretary David Shulkin. Former Environmental Protection Agency Administrator Scott Pruitt also resigned after increasing pressure due to scandals unearthed in public records.

In 2028, we can imagine that many government agencies will have adopted highly sophisticated energy efficiency measures, including smart meters, that can reveal granular information about individual employees' electricity consumption. In fact, this kind of smart-meter data is already being used to gather information on residents; as the *Guardian* reported in 2017, European analytics companies are advertising that smart meters are used to analyze and "build a highly personalized profile for each and every utility customer," and that data is being fed to governments.

The question is when the tables will turn.

For every city, there's always a handful of citizens who wield FOIA requests as a form of retaliation against public officials. In this scenario, a particularly aggravated member of the public files a somewhat abusive request with every city councilmember to learn which offices gobble up the most energy in order to name and shame them during public comment at the next open meeting.

When the records finally come in, he realizes he's hit the motherlode. He figures out when every employee is in or out of the office based on when their light switches and climate controls are toggled on and off. He tabulates how much coffee is being poured on each floor of a government institution, and uncovers when elected officials are using government batteries to charge personal devices unrelated to their jobs. One aide has a gaming console installed in his office, one charges a portable power storage device that they use to power their personal submarine, and, as for Gary Schmitty, he's frequently got a certain remote-controlled sex toy plugged in. It's certainly a safe bet that in 2028, no matter what direction technology goes in, sex scandals will still be a thing.

And then there's one future scenario we can't ignore: a world where, as foreshadowed by the *Washington Post*'s slogan "Democracy Dies in Darkness," open government becomes a relic. Already, the current administration has tossed aside transparency and ethics norms–basic things like releasing tax returns or literally just telling the truth. Across the country, government officials continue to design novel ways of foiling transparency. Sometimes a government agency will file a lawsuit against the citizen who asked for the records. Sometimes they'll dump so many records at once–enough file boxes to build a small fort–that the documents aren't any easier to find than if they'd been shredded.

Some futures are inevitable, but some are avoidable. I believe this is one of those we can steer clear of as long as we acknowledge this one last prediction: we'll need to fight for freedom of information with our pens and teeth today, tomorrow, and every day for the unforeseeable future. ●

JULIA ANGWIN is an award-winning investigative journalist, formerly of the independent news organization *ProPublica* and the *Wall Street Journal*. She has twice led investigative teams that were finalists for a Pulitzer Prize in Explanatory Reporting. In 2003, she was on a team of reporters at the *Wall Street Journal* that was awarded the Pulitzer Prize in Explanatory Reporting for coverage of corporate corruption. Her book *Dragnet Nation: A Quest for Privacy, Security, and Freedom in a World of Relentless Surveillance* was published by Times Books in 2014. She is also the author of *Stealing MySpace: The Battle to Control the Most Popular Website in America* (Random House, March 2009).

MADELINE ASHBY is a science fiction writer and futurist living in Toronto. Her most recent novel, *Company Town*, is available now from Tor Books.

LANDON BATES is a writer from the Central Valley.

ALVARO M. BEDOYA is the founding director of the Center on Privacy and Technology at Georgetown Law, a think tank that studies how government surveillance affects immigrants and people of color. He says things on Twitter at @alvarombedoya.

CINDY COHN is the Executive Director of the Electronic Frontier Foundation. She has spent over twenty-five years as an impact litigator and advocate for the right to have a private conversation online, among other civil liberties in the digital age. The *National Law Journal* named Cohn one of the one hundred most influential lawyers in America in 2013, noting that "if Big Brother is watching, he better look out for Cindy Cohn."

MYKE COLE has a long career in cyber threat intelligence, counterterrorism, and maritime law enforcement with agencies at the federal and local levels, including the U.S. Coast Guard, CIA, DIA, FBI, ONI, and NYPD. He currently performs intelligence and security consulting for the private sector. Myke appeared on CBS's unscripted show *Hunted* last year, where he joined a team of investigators hunting contestants across the southeastern U.S. Cole is also an author of popular fantasy, science fiction, and history. His *Sacred Throne* trilogy is published by Tor, and his *Shadow Ops* and *Reawakening* trilogies by Ace/Roc. His first work of military history, *Legion versus Phalanx*, is out now with Osprey. He has two works of military science fiction, *SAR-1* and *SAR-2*, both coming from Angry Robot this year.

GABRIELLA (BIELLA) COLEMAN holds the Wolfe Chair in Scientific and Technological Literacy at McGill University. She has authored two books, *Coding Freedom: The Ethics and Aesthetics of Hacking* and *Hacker, Hoaxer, Whistleblower, Spy: The Many Faces of Anonymous*.

MALKIA CYRIL is the founder and executive director of the Center for Media Justice and cofounder of the Media Action Grassroots Network, a national network of community-based organizations working to ensure racial and economic justice in a digital age. Cyril is one of few leaders of color in the movement for digital rights and freedom, and a leader in the Black Lives Matter Global Network–helping to bring important technical safeguards and surveillance countermeasures to people across the country who are fighting to reform systemic racism and violence in law enforcement.

CORY DOCTOROW (craphound.com) is a science fiction author, activist, journalist, and the co-editor of Boing Boing (boingboing.net). He is the author of many books, most recently *Walkaway*, a novel for adults; *In Real Life*, a graphic novel; *Information Doesn't Want to Be Free*, a book about earning a living in the internet age; and *Homeland*, a YA sequel to *Little Brother*.

VIRGINIA EUBANKS is an associate professor of Political Science at the University at Albany, SUNY. She is the author of *Automating Inequality: How High-Tech Tools Profile, Police, and Punish the Poor* and *Digital Dead End: Fighting for Social Justice in the Information Age*, and co-editor, with Alethia Jones, of *Ain't Gonna Let Nobody Turn Me Around: Forty Years of Movement Building with Barbara Smith*. Her writing about technology and social justice has appeared in the *American Prospect*, the *Nation*, *Harper's*, and *Wired*. She is a founding member of the Our Data Bodies Project and a Fellow at New America. She lives in Troy, NY.

CAMILLE FASSETT is a reporter at Freedom of the Press Foundation, where she writes about civil liberties and surveillance, and documents attacks on the press in the United States. She is also researcher and resident "public records witch" with

Lucy Parsons Labs, a data liberation and digital security collective.

REYHAN HARMANCI is a writer and editor who lives in Brooklyn.

CHELSEA HOGUE is a writer and educator based in Philadelphia, PA. She wrote an episode for *The Organist* podcast about risk assessment algorithms in criminal justice.

JOANNA HOWARD is the author of *Foreign Correspondent* (Counterpath, 2013), *On the Winding Stair* (BOA Editions, 2009), *In the Colorless Round* (Noemi, 2006), and *Field Glass*, a speculative novel co-written with Joanna Ruocco (Sidebrow, 2017). Her memoir *Rerun Era* is forthcoming from McSweeney's (2019). She lives and works in Providence, RI, and in Denver, CO.

JENNIFER KABAT has been awarded a Creative Capital/Warhol Foundation Arts Writers Grant for her criticism. Her writing has appeared in *Granta*, *Harper's*, *BOMB*, the *Believer*, and the *White Review*. Her essay "Rain Like Cotton" is in *Best American Essays 2018*, and she's working on a book, *Ghostlands*, about grief, modernism, and progressive uprisings as she rebuilds her parents' glass house in rural America. She teaches at the New School and NYU and is a member of her local fire department. Her blue obsession grew out of a meditation on the color commissioned by the Kunsthalle Wien for its exhibition Blue Times.

HAMID KHAN is an organizer with the Stop LAPD Spying Coalition. The mission of the coalition is to build community-based power to dismantle police surveillance, spying, and infiltration programs. The coalition utilizes multiple campaigns to advance an innovative organizing model that is Los Angeles-based but has implications regionally, nationally, and internationally.

ED LOOMIS is a retired NSA computer scientist having worked his entire thirty-seven-year career as a civil servant and an additional five years as a systems architect contractor. He appeared in the recent documentaries *United States of Secrets* and *A Good American* and has published an Amazon memoir of his government experience, *NSA's Transformation: An Executive Branch Black Eye*.

DAVE MAASS is a senior investigative researcher at the Electronic Frontier Foundation and an independent journalist. In 2017, he received the First Amendment Coalition's Free Speech and Open Government Award for his work on surveillance technology. He collects flashlights and lives in San Francisco.

CARSON MELL is an ape on Earth. Some of his rock piles have been regarded highly by other apes. Most have been ignored.

KEN MONTENEGRO is a technologist and lawyer born, raised, and rooted in Los Angeles. He uses what he has learned to advance social change inclined towards mutual aid and liberation. He is the Technology Director at a large civil rights organization, National Vice President of the National Lawyers Guild, and a board member of the Nonprofit Technology (NTEN) and of the Immigrant Defenders Law Center (ImmDef).

JENNY ODELL is an artist and writer based in Oakland, CA. Her writing has appeared in the *New York Times*, *Sierra Magazine*, SFMOMA's *Open Space*, and *Topic*. She has also been an artist-in-residence at Recology SF (otherwise known as the dump), the San Francisco Planning Department, and the Internet Archive. She teaches internet art at Stanford University. Her book *How to Do Nothing* is forthcoming from Melville House.

SORAYA OKUDA is a designer who works on educational materials for beginners and targeted communities. Soraya led the development of EFF's Security Education Companion (sec.eff.org), a curriculum resource for people tasked with teaching digital security to their friends and neighbors.

TREVOR PAGLEN is an artist.

DOUGLAS RUSHKOFF is a media theorist whose books include *Media Virus*, *Program or Be Programmed*, *Present Shock*, and *Throwing Rocks at the Google Bus*. He made the PBS *Frontline* documentaries *Merchants of Cool* and *Generation Like*, and hosts the Team Human radio show and podcast, to be published as a manifesto in January. Portions of his essay were adapted from "The Biology of Disinformation: Memes, Media Viruses, and Cultural Inoculation," by Douglas Rushkoff, David Pescovitz, and Jake Dunagan. (Institute for the Future, March 2018.)

BRUCE SCHNEIER is an internationally renowned security technologist. He teaches at the Harvard Kennedy School, and serves as special advisor to IBM Security. His influential newsletter, Crypto-Gram, and his blog, Schneier on Security, are read by over 250,000 people. His new book is called *Click Here to Kill Everybody: Security and Survival in a Hyper-Connected World*.

JACOB SILVERMAN is the author of *Terms of Service: Social Media and the Price of Constant Connection* (HarperCollins). A contributing editor for the *Baffler*, he has written about technology, politics, and culture for the *New Republic*, the *New York Times*, the *Los Angeles Times*, and many other publications. His website is www. jacobsilverman.com.

EDWARD SNOWDEN is a former intelligence officer who served the CIA, NSA, and DIA for nearly a decade as a subject-matter expert on technology and cybersecurity. In 2013, he provided evidence to journalists that the NSA had established a global mass surveillance network, leading to the most significant reforms to U.S. surveillance policy since 1978. He is the president of the board of the Freedom of the Press Foundation, where he works on methods of enforcing human rights through the development of new technologies.

THENMOZHI SOUNDARARAJAN is a Dalit American transmedia technologist, artist, and organizer who believes story is the most important unit of social change. She is the founder of Equality Lab, a South Asian American tech justice organizing project working at the intersection of organizing, community-based research, storytelling, technology, and digital security to end caste apartheid, Islamophobia, and religious intolerance. She was part of

the inaugural cohort of the Robert Rauschenberg Foundation's Artist as Activist fellows and is currently an inaugural fellow of the Atlantic Fellowship for Racial Equity. Follow her work on Twitter at @dalitdiva and at equalitylabs.org.

ELIZABETH STIX (BERNSTEIN) is a short-story writer and freelance book editor in Berkeley. Her stories have been published in *Tin House*, the *Los Angeles Times Sunday Magazine*, and elsewhere. She is finishing a novel in stories called *Things I Want Back from You*.

SARA WACHTER-BOETTCHER is the author of *Technically Wrong: Sexist Apps, Biased Algorithms, and Other Threats of Toxic Tech* (W.W. Norton). She's also the principal of Rare Union, a consultancy, and the cohost of *No, You Go*, a podcast about living your best feminist life at work. Find her on Twitter at @sara_ann_marie or at sarawb.com.

JENNA WORTHAM is a writer and a journalist for the *New York Times Magazine*. She is the co-host of the podcast *Still Processing* and an energy worker. She lives in Brooklyn, NY.

BEN WIZNER is the director of the ACLU's Speech, Privacy, and Technology Project. He has litigated cases involving digital surveillance, government watch lists, airport security policies, targeted killing, and torture. Since 2013, he has been Edward Snowden's principal legal advisor.

ETHAN ZUCKERMAN directs the Center for Civic Media at MIT and teaches at the MIT Media Lab.

SMALL BLOWS AGAINST ENCROACHING TOTALITARIANISM
edited by McSweeney's

Gathered here are twenty-two pieces in which powerful voices, from poets and novelists to actors and activists, speak to the crucial importance of taking action in 2018.

INFORMATION DOESN'T WANT TO BE FREE
by Cory Doctorow

"Filled with wisdom and thought experiments and things that will mess with your mind." —Neil Gaiman

THE BOATBUILDER

"A meditation on love, loyalty, and the shared experiences that turn strangers into family."
—Tayari Jones

DANIEL GUMBINER

THE BOATBUILDER
by Daniel Gumbiner

"The Boatbuilder offers a decidedly gentle [...] quietly rewarding window onto the attempted recovery of an American opioid addict." —the *New York Times*

HANNAH VERSUS THE TREE
by Leland de la Durantaye

"*Hannah Versus The Tree* is unlike anything I have ever read—thriller, myth, dream, and poem combined." —James Wood

ALL THAT IS EVIDENT IS SUSPECT

Readings from the Oulipo 1963-2018

OULIPO OULIPO

Edited by Ian Monk & Daniel Levin Becker

ALL THAT IS EVIDENT IS SUSPECT: READINGS FROM THE OULIPO 1963–2018
edited by Ian Monk and Daniel Levin Becker

The first English-language collection to offer a life-size picture of the literary group, the Oulipo, in its historical and contemporary incarnations, and the first in any language to represent all of its members.

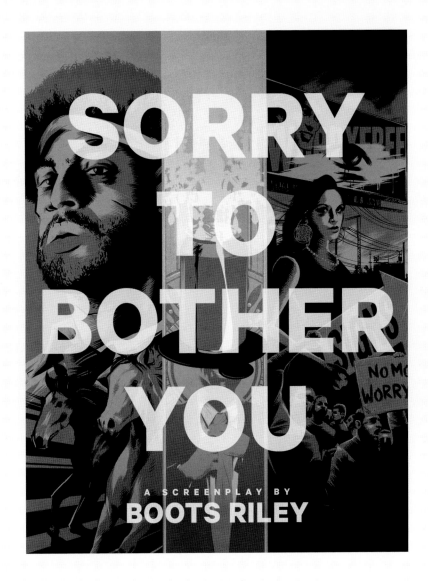

SORRY TO BOTHER YOU
by Boots Riley

Read the original screenplay that sparked the hit major motion picture the *New York Times* called a "must-see" film.

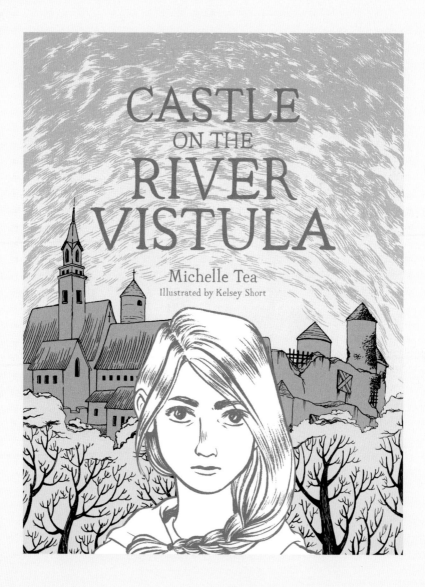

CASTLE ON THE
ON THE
RIVER
VISTULA

Michelle Tea

Illustrated by Kelsey Short

**CASTLE ON THE
RIVER VISTULA**
by Michelle Tea

The final installment of Michelle Tea's
groundbreaking YA adventure series, the
Chelsea Trilogy, here at long last.

IN THE
SHAPE
OF A
HUMAN
BODY
I AM
VISITING
THE
EARTH

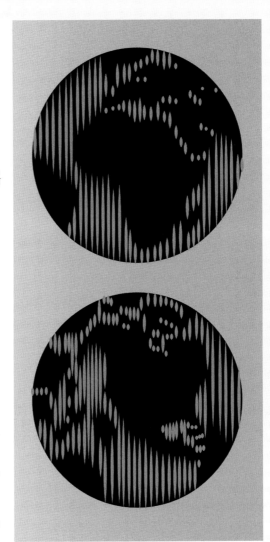

Poems From
Far and Wide

Edited by
Ilya Kaminsky,
Dominic Luxford,
and Jesse Nathan

**IN THE SHAPE OF A HUMAN
BODY I AM VISITING THE EARTH**
Edited by Ilya Kaminsky, Dominic
Luxford, and Jesse Nathan

Poems from far and wide, from the
likes of Pablo Neruda, Federico
García Lorca, Derek Walcott, and
Kwame Dawes

FICTION		
	The Domestic Crusaders	Wajahat Ali
	The Convalescent	Jessica Anthony
	Emmaus	Alessandro Baricco
	Mr. Gwyn	Alessandro Baricco
	All That is Evident is Suspect: Readings from the Oulipo, 1963-2018	Various; Eds. Ian Monk, Daniel Levin Becker
	Arkansas	John Brandon
	Citrus County	John Brandon
	Further Joy	John Brandon
	A Million Heavens	John Brandon
	A Child Again	Robert Coover
	Stepmother	Robert Coover
	One Hundred Apocalypses and Other Apocalypses	Lucy Corin
	Fever Chart	Bill Cotter
	The Parallel Apartments	Bill Cotter
	Sorry to Disrupt the Peace	Patty Yumi Cottrell
	End of I.	Stephen Dixon
	I.	Stephen Dixon
	Hannah Versus the Tree	Leland de la Durantaye
	A Hologram for the King	Dave Eggers
	How We Are Hungry	Dave Eggers
	Understanding the Sky	Dave Eggers
	The Wild Things	Dave Eggers
	You Shall Know Our Velocity	Dave Eggers
	Donald	Stephen Elliott, Eric Martin
	The Pharmacist's Mate and 8	Amy Fusselman
	Painted Cities	Alexai Galaviz-Budziszewski
	The Boatbuilder	Daniel Gumbiner
	God Says No	James Hannaham
	The Middle Stories	Sheila Heti
	Songbook	Nick Hornby
	Bowl of Cherries	Millard Kaufman
	Misadventure	Millard Kaufman
	Lemon	Lawrence Krauser
	Search Sweet Country	Kojo Laing

	Hot Pink	Adam Levin
	The Instructions	Adam Levin
	The Facts of Winter	Paul Poissel
	Sorry to Bother You	Boots Riley
	Adios, Cowboy	Olja Savičević
	A Moment in the Sun	John Sayles
	Between Heaven and Here	Susan Straight
	All My Puny Sorrows	Miriam Toews
	The End of Love	Marcos Giralt Torrente
	Vacation	Deb Olin Unferth
	One Hundred and Forty-Five Stories in a Small Box	Deb Olin Unferth, Sarah Manguso, Dave Eggers
	The Best of McSweeney's	Various
	Noisy Outlaws, Unfriendly Blobs...	Various
	Fine, Fine, Fine, Fine, Fine	Diane Williams
	Vicky Swanky Is a Beauty	Diane Williams
	My Documents	Alejandro Zambra
ART & COMICS	*Song Reader*	Beck
	The Portlandia Activity Book	Carrie Brownstein, Fred Armisen, Jonathan Krisel; Ed. Sam Riley
	The Berliner Ensemble Thanks You All	Marcel Dzama
	It Is Right to Draw Their Fur	Dave Eggers
	Binky Brown Meets the Holy Virgin Mary	Justin Green
	Animals of the Ocean: In Particular the Giant Squid	Dr. and Mr. Doris Haggis-on-Whey
	Children and the Tundra	Dr. and Mr. Doris Haggis-on-Whey
	Cold Fusion	Dr. and Mr. Doris Haggis-on-Whey
	Celebrations of Curious Characters	Ricky Jay
	Be a Nose!	Art Spiegelman
	Everything That Rises: A Book of Convergences	Lawrence Weschler
	826NYC Art Show Catalog	Various
	The Comics Section from the Panorama	Various
BOOKS FOR YOUNGER READERS	*Here Comes the Cat!*	Frank Asch, Vladimir Vagin
	Benny's Brigade	Arthur Bradford; Ill. Lisa Hanawalt
	This Bridge Will Not Be Gray	Dave Eggers; Ill. Tucker Nichols
	The Night Riders	Matt Furie

	The Nosyhood	Tim Lahan
	Hang Glider & Mud Mask	Jason Jägel, Brian McMullen
	Symphony City	Amy Martin
	Crabtree	Jon and Tucker Nichols
	Recipe	Angela and Michaelanne Petrella; Ill. Mike Bertino, Erin Althea
	29 Myths on the Swinster Pharmacy	Lemony Snicket, Lisa Brown
	The Defiant	M. Quint
	The Expeditioners II	S.S. Taylor; Ill. Katherine Roy
	Castle on the River Vistula	Michelle Tea; Ill. Kelsey Short
	Girl at the Bottom of the Sea	Michelle Tea; Ill. Amanda Verwey
	Mermaid in Chelsea Creek	Michelle Tea; Ill. Jason Polan
	The Goods	Various
NONFICTION	*White Girls*	Hilton Als
	In My Home There Is No More Sorrow	Rick Bass
	Maps and Legends	Michael Chabon
	Real Man Adventures	T Cooper
	Information Doesn't Want to Be Free	Cory Doctorow
	The Pharmacist's Mate and 8	Amy Fusselman
	Toro Bravo: Stories. Recipes. No Bull.	John Gorham, Liz Crain
	The End of War	John Horgan
	It Chooses You	Miranda July
	The End of Major Combat Operations	Nick McDonell
	Small Blows Against Encroaching Totalitarianism, Volume One	Various; Ed. McSweeney's
	Mission Street Food	Anthony Myint, Karen Leibowitz
	At Home on the Range	Margaret Yardley Potter, Elizabeth Gilbert
	That Thing You Do With Your Mouth	David Shields, Samantha Matthews
	More Curious	Sean Wilsey
VOICE OF WITNESS	*Throwing Stones at the Moon: Narratives from Colombians Displaced by Violence*	Eds. Sibylla Brodzinsky, Max Schoening
	Palestine Speaks: Narratives of Life under Occupation	Eds. Mateo Hoke, Cate Malek
	Surviving Justice: America's Wrongfully Convicted and Exonerated	Eds. Dave Eggers, Lola Vollen
	Nowhere to Be Home: Narratives from Survivors of Burma's Military Regime	Eds. Maggie Lemere, Zoë West

	Patriot Acts: Narratives of Post-9/11 Injustice	Ed. Alia Malek
	Hope Deferred: Narratives of Zimbabwean Lives	Eds. Peter Orner, Annie Holmes
	Out of Exile: Narratives from the Abducted and Displaced People of Sudan	Ed. Craig Walzer
HUMOR	*The Secret Language of Sleep*	Amelia Bauer, Evany Thomas
	Comedy by the Numbers	Eric Hoffman, Gary Rudoren
	All Known Metal Bands	Dan Nelson
	A Load of Hooey	Bob Odenkirk
	How to Dress for Every Occasion	The Pope
	The Latke Who Couldn't Stop Screaming	Lemony Snicket, Lisa Brown
	The Best of McSweeney's Internet Tendency	Various; Ed. Chris Monks, John Warner
	I Live Real Close to Where You Used to Live	Various; Ed. Lauren Hall
POETRY	*City of Rivers*	Zubair Ahmed
	Remains	Jesús Castillo
	The Boss	Victoria Chang
	Morning in Serra Mattu: A Nubian Ode	Arif Gamal
	Flowers of Anti-Martyrdom	Dorian Geisler
	Of Lamb	Matthea Harvey; Ill. Amy Jean Porter
	Strangest of Theatres: Poets Writing Across Borders	Eds. Jared Hawkley, Susan Rich, Brian Turner
	The Abridged History of Rainfall	Jay Hopler
	Love, an Index	Rebecca Lindenberg
	In the Shape of a Human Body I Am Visiting the Earth	Various; Eds. Ilya Kaminsky, Dominic Luxford, Jesse Nathan
COLLINS LIBRARY	*Curious Men*	Frank Buckland
	Lunatic at Large	J. Storer Clouston
	The Rector and the Rogue	W. A. Swanberg

By subscribing to McSweeney's, you'll receive four beautifully packaged collections of the most groundbreaking stories we can find, delivered right to your doorstep for $95.

QUARTERLY CONCERN SUBSCRIPTION: $95

Four (4) issues of *McSweeney's Quarterly Concern*

If you're renewing your subscription, enter promo code RESUB at checkout to receive $20 off.

Above: McSweeney's 53, an issue disguised as a bag of balloons.

To receive four (4) issues of McSweeney's for $95,
fill out the form below and mail it to:

McSweeney's Subscriptions
849 Valencia Street, San Francisco, CA 94110

Or, subcribe online by navigating to store.mcsweeneys.net.

BILLING NAME: ....

BILLING ADDRESS: ....

CITY: ....

STATE: .... ZIP: ....

EMAIL: ....

PHONE #: ....

CC#: ....

EXPIRES: .... CVV: ....

□ *Check here if billing and mailing address are the same.* □ *Check here if this is a renewal.*

MAILING NAME: ....

MAILING ADDRESS: ....

CITY: ....

STATE: .... ZIP: ....

Please make checks payable to MCSWEENEY'S. *International subscribers please add $42 for shipping.*

Founded in 1998, McSweeney's is an independent publisher based in San Francisco. McSweeney's exists to champion ambitious and inspired new writing, and to challenge conventional expectations about where it's found, how it looks, and who participates. We're here to discover things we love, help them find their most resplendent form, and place them into the hands of curious, engaged readers.

THERE ARE SEVERAL WAYS TO SUPPORT MCSWEENEY'S:

Support Us on Patreon
visit *www.patreon.com/mcsweeneysinternettendency*

Subscribe & Shop
visit *store.mcsweeneys.net*

Volunteer & Intern
email *eric@mcsweeneys.net*

Sponsor Books & *Quarterlies*
email *press@mcsweeneys.net*

To learn more, please visit *www.mcsweeneys.net* or
contact us at *custservice@mcsweeneys.net* or 415.642.5609.